高等学校土木工程专业国际化人才培养英文系列教材

Professional English for Civil Engineering

土木工程专业英语

Editors in Chief

Junwei Liu　Yanchun Liu　Xiaoyang Liu

刘俊伟　刘延春　刘晓阳　主　编

Deputy Editors in Chief

Yuxiang Zhang　Yifei Cui　Yunxia Xia

张玉香　崔祎菲　夏云霞　副主编

中国建筑工业出版社

CHINA ARCHITECTURE & BUILDING PRESS

图书在版编目（CIP）数据

土木工程专业英语＝Professional English for Civil Engineering / 刘俊伟，刘延春，刘晓阳主编；张玉香，崔祎菲，夏云霞副主编．—北京：中国建筑工业出版社，2023.10（2024.9 重印）
高等学校土木工程专业国际化人才培养英文系列教材
ISBN 978-7-112-29130-4

Ⅰ.①土… Ⅱ.①刘… ②刘… ③刘… ④张… ⑤崔… ⑥夏… Ⅲ.①土木工程—英语—高等学校—教材 Ⅳ.①TU

中国国家版本馆CIP数据核字（2023）第172247号

责任编辑：毕凤鸣　齐庆梅　吉万旺
书籍设计：锋尚设计
责任校对：刘梦然
校对整理：张辰双

高等学校土木工程专业国际化人才培养英文系列教材
Professional English for Civil Engineering
土木工程专业英语
Editors in Chief
Junwei Liu Yanchun Liu Xiaoyang Liu
刘俊伟　刘延春　刘晓阳　主　编
Deputy Editors in Chief
Yuxiang Zhang Yifei Cui Yunxia Xia
张玉香　崔祎菲　夏云霞　副主编
*
中国建筑工业出版社出版、发行（北京海淀三里河路9号）
各地新华书店、建筑书店经销
北京锋尚制版有限公司制版
建工社（河北）印刷有限公司印刷
*
开本：787 毫米 × 1092 毫米　1/16　印张：7　字数：138 千字
2023 年 12 月第一版　2024 年 9 月第二次印刷
定价：**36.00** 元（赠教师课件）
ISBN 978-7-112-29130-4
（41847）

"Professional English for Civil Engineering" is a core course for undergraduate students majoring in civil engineering at universities, serving as a crucial foundation for the internationalization of talent in this field. With a focus on cultivating internationally adept engineering professionals, this textbook carefully selects materials that address the practical needs of students seeking opportunities abroad for work or further studies. The content selection strikes a balance between emphasizing core professional concepts and incorporating cutting-edge knowledge.

Comprising nine comprehensive chapters, the textbook covers a range of topics, including Introduction to Civil Engineering, Civil Engineering Materials, Reinforced Concrete Structures, Steel Structures, Bridge Structures, Pavement, Geotechnical Engineering, Geotechnical Investigation, and Project Management. Each chapter is designed to include English course texts, reading materials, summaries, vocabulary, and relevant post-lesson exercises. The material meticulously organizes commonly used professional English vocabulary to help students acquire a solid grasp of essential terminology. The objective is to train students in reading and translating English materials within their field, fostering familiarity with professional English expressions and conventions, ultimately enhancing their English communication skills.

Tailored for use as a professional English textbook, this resource caters to both undergraduate and graduate students majoring in civil engineering. Additionally, it serves as a valuable self-study tool for professionals in the field.

《土木工程专业英语》是高等学校土木工程类专业本科生的一门专业基础课，是土木工程类专业开展国际化人才培养的重要支撑课程。本教材立足于培养国际化高级工程技术人才的目标，结合学生出国工作和进修学习的实际需求，在选材上做了认真的筛选，既注重专业基础内容又强调前沿专业知识。

本教材共九个章节，具体编排如下：土木工程概论、土木工程材料、钢筋混凝土结构、钢结构、桥梁结构、路面工程、岩土工程、岩土勘察、工程项目管理。本书每个章节包括英文课文、英文阅读材料、专业词汇总结及拓展、以及相关课后习题。本教材对常用的专业英语词汇进行了梳理，帮助学生掌握一定数量的专业词汇，训练学生阅读与翻译本专业英文资料的能力，熟悉专业英语的表达方式和表达习惯，提高学生的英语沟通能力。

本教材可作为高校土木工程专业本科生、研究生的专业英语教材，也可以作为相关从业者的自学教材。

本书配备教学课件，可以向采用本书作为教材的老师提供。请有需要的任课教师按以下方式索取课件：1. 邮件：47364196@qq.com 或 jiangongkejian@163.com（邮件主题请注明《土木工程专业英语》英文版）；2. 电话：（010）58337170；3. 建工书院：http://edu.cabplink.com。

序

科学是人类发展之基石，土木工程学则是科学中显学之一。早在学生时代，我便与翻开此书的同学们一样，对土木工程专业外语情有独钟、饶有兴趣，这是因为它富有知识性，且实用性极强。

此书是依托青岛理工大学中美合作办学项目编写完成的，早年我在理工大学工作之际，便有了鼓励老师们编著英文系列教材的想法，感谢老师们的辛苦努力，让此书顺利完成。本书契合了我对一本完美教材的期许，读来收获良多，最重要的是，此书将土木工程专业与英语相结合，可以帮助同学们实现从英语学习到英语应用的过渡。希望本书能够成为理工大学中外合作办学项目学生的良师益友，助力他们早日成为国际化高级工程技术人才！

所谓“单丝不成线，独木不成林”，我希望同学们在学习中能够举一反三，融会贯通，不囿于一种知识，做全面发展的人才，这也是我推荐此书的一大原因。本书不仅可以为中外合作办学项目的同学学习使用，还可以为非中外合作办学项目的学生以及土木工程专业的教师和工程技术人员参考、使用。

“学之广在于不倦，不倦在于固志”，此书中的专业知识不乏晦涩难懂之处，英语的结合更是对同学们学习毅力的一大考验。凡事勤则易，惰则难，这本书的优点不仅在于知识本身，更在于读罢掩卷，同学们的才干、韧劲、毅力，都会有所增长。

当今世界是一个多元化的世界，中国更是一个开放包容的国家，我期望学子们能走出去，面向世界，做开放的、严谨的、科学的、进步的人，所以英语的学习与专业知识一样尤为重要。我可爱的同学们，翻开书吧！博学之、审问之、慎思之、明辨之、笃行之，这不只是我对大家的要求，更是国家寄予同学们的厚望。

同学们能用心阅读是我最大的期待，愿本书能成为大家学习上的好助手！

于德湖

2023 年 8 月

前言

本书依托青岛理工大学中美合作办学项目，根据《普通高等学校本科专业类教学质量国家标准》《普通高等学校教材管理办法》以及《中华人民共和国高等教育法》的基本理念和要求编写完成。坚持准确性、整体性、科学性原则，编排体系既有自己的特点，又突出了和谐统一。本教材面向土木工程专业本科学生，也可作为研究生和土木工程相关从业人员学习参考。

本教材立足于培养国际化高级工程技术人才的目标，结合学生出国工作和进修学习的实际需求，在选材上做了认真的筛选，既注重专业基础内容又强调前沿专业知识。本教材在内容编排上遵循高等教学的基本规律，本着基础性、创新性、适用性的原则，倡导体验、参与、合作与交流的学习方式。顺应土木工程行业发展趋势，紧密结合本科生阶段学生的认知水平，发展学生实际运用、拓展创新和英语思维的能力。

本教材是所有作者共同努力的成果，共九个章节，具体编排如下：土木工程概论、土木工程材料、钢筋混凝土结构、钢结构、桥梁结构、路面工程、岩土工程、岩土勘察、工程项目管理。本书每个章节包括英文课文、英文阅读材料、专业词汇总结及拓展以及相关课后习题。

本教材由青岛理工大学土木工程学院组织编写，感谢各位主编和副主编在教材编写过程中所付出的努力，感谢吕健老师和梁作栋老师为相关章节的编写所付出的辛勤劳动，另外，特别感谢张悦老师为本教材提供封面照片。

本教材虽经多次讨论、统稿，但疏漏乃至错误之处在所难免，望各位读者不吝指正。愿本书能成为大家学习上的好助手!

刘俊伟

2023 年 8 月

Contents

Lesson 1 : Introduction to Civil Engineering

Civil engineering, the oldest of the engineering specialties, is the planning, design, construction, and management of the built environment. This environment includes all structures built according to scientific principles, from irrigation and drainage systems to rocket-launching facilities.

Civil engineers build roads, bridges, tunnels, dams, harbors, power plants, water and sewage system, hospitals, schools, mass transit and other public facilities essential to modern society and large population concentrations. They also build privately owned facilities such as airports, railroads, pipelines, skyscrapers and other large structures designed for industrial, commercial, or residential use. In addition, civil engineers plan, design and build complete cities and towns, and more recently they have been planning and designing space platforms to house self-contained communities.

The word "civil" derives from the Latin for citizen. In 1782, an Englishman John Smeaton used this term to differentiate his nonmilitary engineering work from that of the military engineers who predominated at that time. Since then, the term "civil engineering" has often been used to refer to engineers who build public facilities, although the field is much broader.

1. Scope

Because it is so broad, civil engineering is subdivided into a number of technical specialties. Depending on the type of project, the skills of many kinds of civil engineer specialists may be needed.

When a project begins, the site is surveyed and mapped by civil engineers who locate utility placement water, sewer and power lines. Geotechnical specialists perform soil experiments to determine if the earth can bear the weight of the project. Environmental specialists study the project' s impact on the local area: the potential for air and groundwater pollution, the project' s impact on local animal and plant life, and how the project can be designed to meet government requirements aimed at protecting the environment. Transportation specialists determine what kind of facilities is needed to ease the burden on local roads and other transportation networks that will result from the completed project. Meanwhile, structural specialists use preliminary data to make detailed designs, plans and specifications for the project. Supervising and coordinating the work of these civil engineer specialists, from the beginning to the end of the project, are the construction management specialists. Based on information supplied by the other specialists, construction management civil engineers could estimate quantities and costs of materials and labor, schedule all work, order materials and equipment for the job, hire contractors and subcontractors, and perform other supervisory work to ensure the project is completed on time and as specified.

Throughout any given project, civil engineers make extensive use of computers. Computers are used to design the project' s various elements (computer-aided design, or CAD) and to manage it. Computers are a necessity for the modern civil engineer because they permit the engineer to efficiently handle the large quantities of data needed in determining the best way to construct a project.

2. Structural engineering

In this specialty, civil engineers plan and design structures of all types, including bridges, dams, power plants, supports of equipment, special structures for offshore projects, the United States space program, transmission towers, giant astronomical and radio telescopes, and many other kinds of projects. By using computers, structural engineers determine the forces a structure must resist: its own weight, wind and hurricane forces, temperature changes that expand or contract construction materials, and earthquakes. They also determine the combination of appropriated materials, steel, concrete, plastic, stone, asphalt, brick, aluminum, or other construction materials.

3. Water resources engineering

Civil engineers in this specialty deal with all aspects of the physical control of water. Their projects help prevent floods, supply water for cities and for irrigation, manage and control rivers and water runoff, and maintain beaches and other waterfront facilities. In addition, they also design and maintain harbors, canals and locks, build huge hydroelectric dams and smaller dams and water impoundments of all kinds, help design offshore structures, and determine the location of structures affecting navigation.

4. Geotechnical engineering

Civil engineers who specialize in this field analyze the properties of soils and rocks that support structures and affect structural behavior. They evaluate and work to minimize the potential settlement of buildings and other structures that stems from the pressure of their weight on the earth. These engineers also evaluate and determine how to strengthen the stability of slopes and fills and how to protect structures against earthquakes and the effects of groundwater.

5. Environmental engineering

In this branch of engineering, civil engineers design, build and supervise systems to provide safe drinking water and to prevent and control the pollution of water supplies, both on the surface and underground. They also design, build and supervise projects to control or eliminate pollution of the land and air. These engineers build water and wastewater treatment plants, and design air scrubbers and other devices to minimize or eliminate air pollution caused by industrial processes, incineration, or other smoke-producing activities. They also work to control toxic and hazardous wastes through the construction of special dump sites or the neutralizing of toxic and hazardous substances. In addition, the engineers design and manage sanitary landfills to prevent pollution of surrounding land.

6. Transportation engineering

Civil engineers working in this specialty build facilities to ensure safe and efficient movements of both people and goods. They specialize in designing and maintaining all types of transportation facilities, highways and streets, mass transit systems, railroads and airfields, ports and harbors. Transportation engineers apply technological knowledge as well as consideration of the economic, political and social factors in designing each project. They work closely with urban planners, since the quality of the community is directly related to the quality of the transportation system.

7. Pipeline engineering

In this branch of civil engineering, engineers build pipelines and related facilities which transport liquids, gases, or solids ranging from coal slurries (mixed coal and water) and semi-liquid wastes, to water, oil, and various types of highly combustible and noncombustible gases. The engineers determine pipeline design, the economic and environmental impact of a project on regions it must traverse, the type of materials to be used — steel, concrete, plastic, or combinations of various materials — installation techniques, methods for testing pipeline strength, and controls for maintaining proper pressure and rate of flow of materials being transported. When hazardous materials are being carried, safety is a major consideration as well.

8. Construction engineering

Civil engineers in this field oversee the construction of a project from beginning to end. Sometimes called project engineers, they apply both technical and managerial skills, including knowledge of construction methods, planning, organizing, financing and operating construction projects. They coordinate the activities of virtually everyone engaged in the work: the surveyors, workers who lay out and construct the temporary roads and ramps, excavate for the foundation, build the forms and pour the concrete, and workers who build the steel framework. These engineers also make regular progress reports to the owners of the structure.

9. Community and urban planning

Those engaged in this area of civil engineering may plan and develop communities within a city, or entire cities. Such planning involves far more than engineering consideration; environmental, social and economic factors in the use and development of land and natural resources are also key elements. These civil engineers coordinate planning of public works along with private development. They evaluate the kinds of facilities needed, including streets and highways, public transportation systems, airports, port facilities, water-supply and waste water-disposal systems, public buildings, parks, and recreational and other facilities to ensure social and economic as well as environmental well-being.

10. Photogrametry, surveying and mapping

The civil engineers in this specialty precisely measure the Earth' s surface to obtain reliable information for locating and designing engineering projects. This practice often involves high-technology methods such as satellite and aerial surveying, and computer processing of photographic imagery. Radio signals from satellites, scans by laser and sonic beams, are converted to maps to provide far more accurate measurements for boring tunnels, building highways and dams, plotting flood control and irrigation projects, locating subsurface geologic formations that may affect a construction project, and a host of other building uses.

11. Other specialties

Two additional civil engineering specialties that are not entirely within the scope of civil engineering but are essential to the discipline are engineering management and engineering teaching.

(1) Engineering management

Many civil engineers prefer careers that eventually lead to management. Others are able to start their careers in management positions. The civil engineer-manager combines technical knowledge with the ability to organize and coordinate worker power, materials, machinery and funds. These engineers may work in government — municipal, county, state, or federal. In the U. S. Army Corps of Engineers they work as military or civilian management engineers; or in semiautonomous regional or city authorities or similar organizations. They may also manage private engineering firms ranging in scale from a few employees to hundreds.

(2) Engineering teaching

The civil engineers who prefer a teaching career usually teach both graduate and undergraduate students in technical specialties. Many teaching civil engineers engage in basic research that eventually leads to technical innovations in construction materials and methods. Many also serve as consultants on engineering projects, or on technical boards and commissions associated with major projects.

1.1 Phrases and Expressions

irrigation |ˌɪrəˈgeɪʃən| *n.* 灌溉
drainage |ˈdreɪnɪdʒ| *n.* 排水
sewage |ˈsuːɪdʒ| *n.* 污水；下水道；污物
mass transit |mæs ˈtrænsɪt| 公共交通；大量客运
predominate |prɪˈdɑːmɪneɪt| *vi.* 占支配地位；（数量上）占优势
survey |ˈsɜːrveɪ, sərˈveɪ| *n.* 勘测；*v.* 测量
geotechnical |dʒiːəuˈteknikəl| *adj.* 岩土工程技术的
transmission tower |trænsˈmɪʃən ˈtaʊɚ| 输电塔
astronomical |ˌæstrəˈnɑːmɪkl| *adj.* 天文学的
hurricane |ˈhɜːrəkeɪn| *n.* 飓风
asphalt |ˈæsfɔlt| *n.* 沥青
impoundment |ɪmˈpaʊndmənt| *n.* 蓄水；积水
navigation |ˌnævɪˈgeɪʃn| *n.* 航海
settlement |ˈsetlmənt| *n.* 下沉；沉降量
scrubber |ˈskrʌbɚ| *n.* 刷子；清洗者
incineration |ɪnˌsɪnəˈreɪʃn| *n.* 焚化
hazardous |ˈhæzərdəs| *adj.* 危险的；有害的
dump |dʌmp| *v.* 丢弃，乱堆；*n.* 废料堆场

neutralize |ˈnuːtrəlaɪz| *vt.* 使中立；使无效；中和
sanitary |ˈsænəteri| *adj.* 清洁的；公共卫生的
slurry |ˈslɜri| *n.* 泥浆，浆体
combustible |kəmˈbʌstəbəl| *adj.* 易燃的；可燃的；*n.* 可燃物；易燃物
ramp |ræmp| *n.* 斜坡；坡道
photogrametry |foʊtəuˈgræmɪtrɪ| *n.* 摄影测量
aerial |ˈeriəl| *adj.* 空中的；*n.* 天线
sonic |ˈsɑnɪk| *adj.* 音速的；声音的
plot |plɑːt| *n.* 图表；*v.* 绘制
municipal |mjuːˈnɪsɪpl| *adj.* 城市的；市政的

1.2 Please translate the following sentences into Chinese

(1) Civil engineers build roads, bridges, tunnels, dams, harbors, power plants, water and sewage systems, hospitals, schools, mass transit, and other public facilities essential to modern society and large population concentrations.

(2) Based on information supplied by the other specialists, construction management civil engineers estimate quantities and costs of materials and labor, schedule all work, order materials and equipment for the job, hire contractors and subcontractors, and perform other supervisory work to ensure the project is completed on time and as specified.

(3) By using computers, structural engineers determine the forces a structure must resist: its own weight, wind and hurricane forces, temperature changes that expand or contract construction materials, and earthquakes.

(4) They evaluate and work to minimize the potential settlement of buildings and other structures that stems from the pressure of their weight on the earth.

(5) They coordinate the activities of virtually everyone engaged in the work: the surveyors, workers who lay out and construct the temporary roads and ramps, excavate

for the foundation, build the forms and pour the concrete, and workers who build the steel framework.

1.3 Please translate the following sentences into English

（1）在设计任何土木工程项目之前，必须先对建筑场地进行测量。

（2）工业建筑用于各种工厂或工业生产，而民用建筑指的是那些人们用以居住、工作、教育或进行其他社会活动的场所。

（3）均匀沉降不会产生很严重的后果，但是不均匀沉降有破坏性影响，比如，建筑物可能倾斜，墙体可能开裂，门窗可能被破坏。

（4）钢材比混凝土具有更高的抗拉和抗压性能，所以钢结构建筑物重量更轻。

（5）结构设计是选择材料和构件的类型、尺寸和形状，以安全耐久的方式来承担荷载。

1.4 Extension

Infrastructure |ˈɪnfrəstrʌktʃər| *n.* 基础设施
mechanical system |mɪˈkænɪkəl ˈsɪstəm| 机械系统，力学系统
thermal insulation |ˈθɚməl ˌɪnsəˈleʃən| 绝热；保温
sustainable |səˈsteɪnəbl| *adj.* 可持续的
aesthetic |esˈθetɪk| *adj.* 美学的；美的；*n.* 美感；美学
procurement |prəˈkjʊrmənt| *n.* 采购；获得
budget |ˈbʌdʒɪt| *n.* 预算；*v.* 制定预算；计划；*adj.* 低价的
fire endurance |faɪrɛnˈdʊrəns| 耐火性；耐火极限
ventilation |ˌvɛntəˈleɪʃən| *n.* 通风
conservation |ˌkɑːnsərˈveɪʃn| *n.* 保护；节约
building acoustics |ˈbildiŋəˈkuːstɪks| 建筑声学
prefabrication |ˌprifæbrɪˈkeɪʃn| *n.* 预制
scaffold |ˈskæfoʊld| *n.* 脚手架
retaining wall |rɪˈtenɪŋwɔːl| 挡土墙，护岸
seismic load |ˈsaɪzmɪklod| 地震荷载
durability |ˌdjʊrəˈbɪlətɪ| *n.* 耐用性

ecological |ˌiːkəˈlɑːdʒɪkl| *adj.* 生态（学）的
illumination |ɪˌluməˈneʃən| *n.* 照明

1.5 Reading material

Performance requirements of the building fabric

The requirement to provide an acceptable internal environment is simply one of the performance requirements of modern buildings. The level of performance of buildings depends upon several factors, the emphasis which is placed upon these individual performance requirements varies from situation to situation. However, minimum standards are set out by statutes and guidelines, such as the Building Regulations, which must be achieved in any instance. The increasing role of the building as an asset has also affected the ways in which buildings have been designed to maximise the long-term value and minimise the maintenance costs of the structure and fabric.

The performance requirements of buildings may be summarised as follows.

Structural stability

In order to satisfactorily fulfil the functions of it, a building must be able to withstand the loadings imposed upon it without suffering deformation or collapse. This necessitates the effective resistance of loadings or their transfer through the structure to the ground.

Durability

The long-term performance of the structure and fabric requires that the component parts of the building are able to withstand the vagaries and hostilities of the environment in which they are placed, without deterioration. The ability of the parts of the building to maintain their integrity and functional ability for the required period of time is fundamental to the ability of the building to perform in the long term. This factor is particularly affected by the occurrence of fires in buildings.

Thermal insulation

The need to maintain internal conditions within fixed parameters and to conserve energy dictate that the external fabric of a given building provides an acceptable standard of resistance to the passage of heat. The level of thermal insulation which is desirable in an individual instance is, of course, dependent upon the use of the building, its location and so on.

Exclusion of moisture and protection from weather

The passage of moisture from the exterior, whether in the form of ground water rising through capillary action, precipitation or other possible sources, should be resisted by the building envelope. The ingress of moisture to the building interior can have several undesirable effects, such as the decay of timber elements, deterioration of surface finishes and decorations and risks to health of occupants, in addition to effects upon certain processes carried out in the building. Hence details must be incorporated to resist the passage of moisture, from all undesirable sources, to the interior of the building. The exclusion of wind and water is essential to the satisfactory performance of any building fabric.

Acoustic insulation

The passage of sound from the exterior to the interior, or between interior spaces, should be considered in building construction. The level of sound transmission which is acceptable in a building will vary considerably, depending upon the nature of the use of the building and its position.

Flexibility

In industrial and commercial buildings in particular, the ability of the building to cope with and respond to changing user needs has become very important. Hence the level of required future flexibility must be taken into account in the initial design of the building. this is reflected, for example, in the trend to create buildings with large open spaces, which may be subdivided by the use of partitions which may be readily removed and relocated.

Aesthetics

The issue of building aesthetics is subjective. However, it should be noted that in some situations the importance of the building' s aesthetics is minimal, while in others, of course, it is highly important.

For example, the appearance of a unit on an industrial estate is far less important than that of a city centre municipal building. The extent to which aesthetics are pursued will have an inevitable effect on the cost of the building.

This summary is not a definitive list of the performance requirements of all building components in all situations. However, it is indicative of the factors which affect the design and performance of buildings and their component parts.

课文选自：宿晓萍，赵庆明. 土木工程专业英语 [M]. 北京：北京大学出版社，2017.

阅读选自：Mike Riley and Alison Cotgrave. Industrial and commercial building[Z]. 2004.

Lesson 2 :
Civil Engineering
Materials

The earth is composed of various materials. Materials science and engineering serves as the base for all technology branches such as electronics, energy, communication, environment, and health engineering. Construction materials are the most widely used materials and their usage is the largest in terms of tonnage in the world.

Throughout the history of human civilization, many kinds of materials such as clay, brick, rock, concrete, wood and steel have been used in the construction of buildings, bridges, roads and other structures. According to their load-carrying function in structures, construction materials can be divided into structural materials and non-structural materials. Structural materials carry not only their own self-weight, but also loads being transferred from other components of buildings or infrastructures. On the other hand, non-structural materials only carry their own weight. Non-structural materials include floor and wall coverings, tiles, glass insulation materials, sealants and paint or coatings. Most of them are specified for protection, aesthetic and architectural purposes by the architect or the interior designer.

This chapter focuses on modern structural materials including concrete, steel, wood, bituminous materials as well as polymers and fibrous composites. Among these materials, concrete will receive the most attention for two reasons. First, the civil engineer is responsible for designing the concrete he/she uses and for ensuring its long term performance. It should be pointed out that steel and wood products are designed by material and mechanical engineers, who supply them to civil engineers according to their specifications. Second, concrete (reinforced concrete) is the most widely-used construction material in the world.

Besides concrete, steel and wood are the other two most commonly-used construction materials in the world. In the US, for example, most residential houses are built with wood and over half of the office buildings are constructed with steel. This is due to the abundant supply of both materials, making them economical. Steel, besides its use as structural members on its own, is also used as reinforcements or prestressed tendons for concrete structures. Understanding steel behavior is hence an important component in the study of reinforced concrete and prestressed concrete design. Bituminous materials are used all over the world in the construction of road pavements. In recent years, polymers and polymeric composites have been gaining popularity in the construction industry due to their light weight and good durability. Polymers have been used in pipes, fabrics for roofing as well as geotextiles for slope protection. Reinforcing bars and gribs have been made with fiber reinforced composites to replace metals in corrosive environments. Moreover, structural renovation in concrete utilizes more and more fibrous composites.

Soil is also an important construction material. Masonry (bricks and blocks) are widely used in building walls. Since they are not the primary load carrying components, they are not discussed here.

1. Phrases and expressions

materials science *n.* 材料科学
construction materials *n.* 建筑材料
human civilization *n.* 人类文明
concrete*n.* |ˈkɒŋkriːt| *n.* 混凝土
steel |stiːl| *n.* 钢铁
wood |wʊd| *n.* 木材
self-weight *n.* 自重
polymer |ˈpɑːlɪmər| *n.* 聚合物
pressed concrete *n.* 预应力混凝土
tiles |taɪlz| *n.* 瓷砖

2. Please translate the following sentences into Chinese

(1) Materials science and engineering serves as the base for all technology

branches such as electronics, energy, communication, environment and health engineering.

(2) Throughout the history of human civilization, many kinds of materials such as clay, brick, rock, concrete, wood and steel have been used in the construction of buildings, bridges, roads and other structures.

(3) Besides concrete, steel and wood are the other two most commonly-used construction materials in the world.

(4) Among these materials, concrete will receive the most attention for two reasons.

(5) Non-structural materials include floor and wall coverings, tiles, glass insulation materials, sealants and paint or coatings.

3. Please translate the following sentences into English

（1）让我们享受舒适生活的各种新技术的发展直接依赖于有合适的新材料。

（2）通常，金属材料具有良好的导电性和导热性，具有金属光泽，密度较大，并具有在荷载下变形而不断裂的能力。

（3）尽管陶瓷与复合材料结合可以显著改善陶瓷的韧性，但是在通常情况下，陶瓷的韧性都比较差。

（4）高聚物是由共价键组成的结构单元不断重复组成的大分子，这个单词是由希腊字母衍生而来的，poly 的意思是很多，meros 的意思是部分的。

（5）由于医学、生物学和材料工程学的进步，不仅我们的寿命更长，而且我们的生活质量也在飞速提高。

4. Extension

durability |dʊrəˈbɪləti| *n.* 耐久性
thermal conductivity *n.* 热导性
portland cement *n.* 硅酸盐水泥

permeability |pɜːrmiəˈbɪləti| *n.* 抗渗性
intermetallic |ˌɪntəmɪˈtælɪk| *adj.* 金属的
compound |ˈkɑːmpaʊnd| *n.* 金属化合物
thermoplastic |ˌθɜːrmoʊˈplæstɪk| *n.* 热塑性塑料
alumina |əˈluːmɪnə| *n.* 铝
polymerization |ˌpɒliməraɪˈzeɪʃn| *n.* 聚合反应
inorganic polymer *n.* 无机聚合物
polypropylene |ˌpɑːliˈproʊpəliːn| *n.* 聚丙烯
low carbon steel *n.* 低碳钢
ceramic & pottery *n.* 陶瓷
interfacial transition zone (ITZ) *n.* 界面过渡区
ductility |dʌkˈtɪləti| *n.* 韧性
alkaline earth metal *n.* 碱土金属

5. Reading material

The durability and green civil material

The wide-ranging green transformation of China in the next five years and beyond, has been regarded as a basic state policy, marks a historic turning point in our country's development mode. The ambitious vision is also of global significance, because it will offer other developing nations an alternate path to prosperity rather than just following in the footsteps of developed countries.

While China is increasingly prepared for the transformation in terms of its technological and economic strength, our country still faces complicated and arduous tasks in hammering out systematic action plans to fulfill the vision. To achieve a fundamental improvement in environmental quality by 2035, China will endeavor to make production and lifestyles green throughout all areas of society, according to a proposal unveiled in November after the Fifth Plenary Session of the 19th Central Committee of the Communist Party of China in October. The country will make efforts to see marked progress in the transformation during the 14th Five-year Plan (2021—2025) period, according to the proposal.

The concept of green development has been an underlying trend since the 18th CPC National Congress in 2012, and the proposal shows that, with stronger governance capabilities, the central authorities have made sustained efforts to further develop Xi Jinping Thought on Ecological Civilization, a concept promoted by Xi that advocates balanced and sustainable development. The Ecological Civilization on civil engineering area promoted the development on durability design and sustainable materials. The research and application of durability problems such as corrosion, carbonation and froze-thaw, as well as green cement geopolymer and so on, are strongly increased.

Green development is one of President Xi Jinping's top concerns. Via video link in Beijing at the Leaders' Side Event on Safeguarding the Planet of the G20 Riyadh Summit on Nov 22, for example, he said China will pursue clean, low-carbon, safe and efficient use of energy and green industries to promote greener economic and social development in all respects. In order to chase the path towards the direction of clean, low-carbon, safe and efficient use of energy and greener engineering, the civil engineering industry should also reduce the energy and environment cost through the whole industrial chain.

Instead of considering GDP as the core focus of development, the central authorities have been tending to prioritize environmental protection in their governance philosophy, or at least emphasizing the need to balance economic development and environmental protection. For the engineering department, the need is to balance the short and long term bearing capacity and environmental and energy cost.

Previously, China mainly resorted to "end-of-pipe solutions" — pollution-control approaches that clean up pollutants at the point where they enter the environment — in its environmental governance. Now it has been shifting to a more systematic approach of reforming the economic development mode by adjusting the structures of industry, energy consumption, transportation and land use.

For the civil engineering field, which is one of the most crucial part in the 2035 environmental plan, has been facing extra challenge on the green transformation. The main problem of civil engineering is the pollution in the building material production and the economic and environmental cost of durability. The pressure from inside and outside the industry are pushing the development of sustainable civil engineering materials and technology. Aside from strengthening law and policy support for green construction materials, for example, it said the country will strive to promote green real estate and encourage green technological innovations, such as the green cement production, the passive house and the sustainable design.

The green transmission in the civil engineering industry further strengthens the research on durability construction materials and technology. The corrosion and protection of reinforced concrete structures, for example, one of the the long-term concerns of all constructors and researchers, has been paid even more attention as a mean contributor to the sustainable growth. Steel corrosion constitutes the most important cause of premature aging and deterioration on reinforced-concrete structures on an international scale. Its technical, financial and societal consequences are considerable.

The marine environment in city such as Qingdao, is one of the worst cases for corrosion and other durability problems. The actual life of marine concrete structures is usually much shorter than the designed service time due to various attacks from seawater. Chloride-induced reinforcement corrosion is regarded as the primary durability issues. Although corrosion resistant reinforcements including stainless steel and fiber-reinforced polymers (FRP) have been proposed, the carbon steel is still irreplaceable currently in field construction due to the practical advantages of low cost, easy field processing, versatile mechanical performances, etc. In reinforced concrete (RC) system, steel reinforcements are chemically protected by alkaline concrete pore solution and physically protected by the barrier effect of the dense concrete materials, which can inhibit the access of aggressive species. That means the resistance of concrete to seawater attack is decisive to service time of RC structure in the marine environment, which is an obstacle for the maintenance of marine structures.

With the progress of green transmission in civil engineering industry, the durability and sustainability of structures becomes a more and more important responsibility of engineers. There is no excuse to let the structures in marine and other severe environment fail before their designed term of service. The durability and actual service life of concrete are affected by improper selection mixture materials or inadequate quality control during construction. For the reinforced concrete structures, it is essential to protect steel reinforcement from corrosion through strict concrete design. Since all aggressive ions penetrate through pore structure, the permeability of concrete is important for durability.

Lesson 3 :
Reinforced Concrete Structures

Reinforced concrete is concrete in which reinforcement bars, reinforcement grids, plates or fibers have been incorporated to strengthen the concrete in tension. It was invented by French gardener Joseph Monier in 1849 and patented in 1867. The term Ferro Concrete only refers to concrete that is reinforced with iron or steel. Other materials used to reinforce concrete can be organic and inorganic fibers as well as composites in different forms. Concrete is strong in compression, but weak in tension, thus adding reinforcement increases the strength in tension. In addition, the failure strain of concrete in tension is so low that the reinforcement has to hold the cracked sections together. For a strong, ductile and durable construction the reinforcement shall have the following properties: high strength, high tensile strain, good bond to the concrete, thermal compatibility, durability in the concrete environment. In most cases reinforced concrete uses steel rebars that have been inserted to addstrength.

Concrete is reinforced to give it extra tensile strength; without reinforcement, many concrete buildings would not exist. Reinforced concrete can encompass many types of structures and components, including slabs, walls, beams, columns, foundations, frames and more. Reinforced concrete can be classified as precast or cast in-situ concrete. Usually we pay more attention to floor systems, But designing and implementing the most efficient floor system is key to creating optimal building structures. Small changes in the design of a floor system may have significant impact on material costs, construction schedule, ultimate strength, operating costs, occupancy levels and end use of a building.

Concrete is a mixture of coarse (stone or brick chips) and fine (generally sand) aggregates with a binder material (usually Portland cement) . When mixed with a small amount of water, the cement hydrates form microscopic opaque crystal lattices encapsulating and locking the aggregate into a rigid structure. Typical concrete mixes have high resistance to compressive stresses (about 4,000 psi (28 MPa)) ; however, any appreciable tension (e.g. due to bending) will break the microscopic rigid lattice, resulting in cracking and separation of the concrete. For this reason, typical non-reinforced concrete must be well supported to prevent the development of tension.

If certain material with high strength in tension, such as steel, is placed in concrete, then the composite material, reinforced concrete, resists not only compression but also bending and other direct tensile actions. A reinforced concrete section where the concrete resists the compression and steel resists the tension can be made into almost any shape and size for the construction industry.

Three physical characteristics give reinforced concrete its special properties. First, the coefficient of thermal expansion of concrete is similar to that of steel, eliminating large internal stresses due to differences in thermal expansion or contraction. Second, when the cement paste within the concrete hardens this conforms to the surface details of the steel, permitting any stress to be transmitted efficiently between the different materials. Usually steel bars are roughened or corrugated to further improve the bond or cohesion between the concrete and steel. Third, the alkaline chemical environment provided by the alkali reserve (KOH, NaOH) and the portlandite (calcium hydroxide) contained in the hardened cement paste causes a passivating film to form on the surface of the steel, making it much more resistant to corrosion than it would be in neutral or acidic conditions. When the cement paste exposed to the air and meteoric water reacts with the atmospheric CO_2, portlandite and the Calcium Silicate Hydrate (CSH) of the hardened cement paste become progressively carbonated and the high pH gradually decreases from 13.5-12.5 to 8.5, the pH of water in equilibrium with calcite (calcium carbonate) and the steel is no longer passivated.

As a rule of thumb, only to give an idea on orders of magnitude, steel is protected at pH above 11 but starts to corrode below 10 depending on steel characteristics and local physico-chemical conditions when concrete becomes carbonated. Carbonation of concrete along with chloride ingress are amongst the chief reasons for the failure of reinforcement bars in concrete.

The relative cross-sectional area of steel required for typical reinforced concrete is usually quite small and varies from 1% for most beams and slabs to 6% for some columns. Reinforcing bars are normally round in cross-section and vary in diameter. Reinforced concrete structures sometimes have provisions such as ventilated hollow cores to control their moisture&humidity.

A beam bends under bending moment, resulting in a small curvature. At the outer face (tensile face) of the curvature the concrete experiences tensile stress, while at the inner face (compressive face) it experiences compressive stress.

A singly-reinforced beam is one in which the concrete element is only reinforced near the tensile face and the reinforcement, called tension steel, is designed to resist the tension. A doubly-reinforced beam is one in which besides the tensile reinforcement the concrete element is also reinforced near the compressive face to help the concrete resist compression. The latter reinforcement is called compression steel. When the compression zone of a concrete is inadequate to resist the compressive moment (positive moment), extra reinforcement has to be provided if the architect limits the dimensions of the section.

An under-reinforced beam is one in which the tension capacity of the tensile reinforcement is smaller than the combined compression capacity of the concrete and the compression steel (under-reinforced at tensile face) . When the reinforced concrete element is subject to increasing bending moment, the tension steel yields while the concrete does not reach its ultimate failure condition. As the tension steel yields and stretches, an "under-reinforced" concrete also yields in a ductile manner, exhibiting a large deformation and warning before its ultimate failure. In this case, the yield stress of the steel governs the design.

An over-reinforced beam is one in which the tension capacity of the tension steel is greater than the combined compression capacity of the concrete and the compression steel (over-reinforced at tensile face) . So the "over-reinforced concrete" beam fails by crushing of the compressive-zone concrete and before the tension zone steel yields, which does not provide any warning before failure as the failure is instantaneous. A balanced-reinforced beam is one in which both the compressive and tensile zones reach yielding at the same imposed load on the beam, and the concrete will crush and the tensile steel will yield at the same time. This design criterion is however as risky as over-reinforced concrete, because failure is sudden as the concrete crushes at the same time of the tensile steel yields, which gives a very little warning of distress in tension failure.

Steel-reinforced concrete moment-carrying elements should normally be designed to be under-reinforced so that users of the structure will receive warning of impending collapse. The characteristic strength is the strength of a material where less than 5% of the specimen shows lower strength. The design strength or nominal strength is the strength of a material, including a material-safety factor. The value of the safety factor generally ranges from 0.75 to 0.85 in Allowable Stress Design. The ultimate limit state is the theoretical failure point with a certain probability. It is stated under factored loads and factored resistances.

Reinforced concrete can fail due to inadequate strength, leading to mechanical failure, or due to a reduction in its durability. Corrosion and freeze/thaw cycles may damage poorly designed or constructed reinforced concrete. When rebar corrodes, the oxidation products (rust) expands and tends to flake, cracking the concrete and unbonding the rebar from the concrete. Typical mechanisms leading to durability problems are discussed below.

Mechanical failure: Cracking of the concrete section can not be prevented. However, the size of and location of the cracks can be limited and controlled by reinforcement, placement of control joints, the curing methodology and the mix design of the concrete. Cracking defects allow moisture to penetrate and corrode the reinforcement. This is a serviceability failure in limit state design. Cracking is normally the result of an inadequate quantity of rebar, or rebar spaced at too great a distance. The concrete then cracks either under excess loading, or due to internal effects such as early thermal shrinkage when it cures.

Ultimate failure leading to collapse can be caused by crushing of the concrete, when compressive stresses exceed its strength; by yielding or failure of the rebar, when bending or shear stresses exceed the strength of the reinforcement; or by bond failure between the concrete and the rebar.

1. Phrases and Expressions

reinforced concrete *n.* 钢筋混凝土
tensile strength *n.* 抗拉强度
reinforcement bar/steel rebar *n.* 钢筋
reinforcement grid *n.* 钢筋网
bond |bɑːnd| *n.* 粘结
slab |slæb| *n.* 板
beam |biːm| *n.* 梁
column |kɑːləm| *n.* 柱
frame |freɪm| *n.* 框架
precast concrete *n.* 预制混凝土
cast in-situ concrete *n.* 现浇混凝土
ultimate strength *n.* 极限强度
binder material *n.* 胶粘剂
cross-sectional area *n.* 界面面积
bending moment *n.* 弯矩
doubly-reinforced beam *n.* 双筋梁
over-reinforced beam *n.* 超筋梁
corrosion |kəˈrəʊʒn| *n.* 侵蚀
freeze/thaw cycles *n.* 冻融循环
crack |kræk| *n.* 裂缝
cover to reinforcement *n.* 钢筋保护层

2. Please translate the following sentences into Chinese

(1) A reinforced concrete section where the concrete resists the compression and steel resists the tension can be made into almost any shape and size for the construction industry.

(2) The coefficient of thermal expansion of concrete is similar to that of steel, eliminating large internal stresses due to differences in thermal expansion or contraction.

(3) Usually steel bars are roughened or corrugated to further improve the bond or cohesion between the concrete and steel.

(4) A doubly-reinforced beam is one in which besides the tensile reinforcement the concrete element is also reinforced near the compressive face to help the concrete resist compression.

(5) Steel-reinforced concrete moment-carrying elements should normally be designed to be under-reinforced so that users of the structure will receive warning of impending collapse.

3. Please translate the following sentences into English

（1）混凝土的抗压强度很高，但抗拉强度却很低，其抗拉强度是其抗压强度的百分之十左右。

（2）在荷载作用下，这个力可以消除跨中关键部位和支座部位的拉应力，从而减小裂缝开展。

（3）在一般的钢筋混凝土结构中，通常认为混凝土的抗拉强度是可以忽略不计的。

（4）为了消除荷载所引起的纯拉应力，在构件承受恒荷载和活荷载前，就预先给他们施加一个永久的预压应力。

（5）使用高强混凝土和钢筋的构件比普通钢筋混凝土构件更轻质。

4. Extension

eccentric |ɪkˈsentrɪk| *adj.* 偏心的
loss of prestress *n.* 预应力损失
plain concrete *n.* 素混凝土
long-term deflection *n.* 长期变形

creep |kri:p| *n.* 徐变
deformed bar *n.* 变形钢筋
stirrup |stɪrəp| *n.* 箍筋
neutral axis *n.* 中性轴
prestressed tendon *n.* 预应力钢丝束
anchorage |ˈæŋkərɪdʒ| *n.* 锚具

5. Reading material

Prestressed Concrete

Concrete is strong in compression, but weak in tension, its tensile strength varies from 8 to 14 percent of its compressive strength. Due to such a low tensile capacity, flexural cracks develop at early stages of loading. In order to reduce or prevent such cracks from developing, a concentric or eccentric force is imposed in the longitudinal direction of the structural element. This force prevents the cracks from developing by eliminating or considerably reducing the tensile stresses at the critical midspan and support sections at service load, thereby raising the bending, shear and torsional capacities of the sections. The sections are then able to behave elastically, and almost the full capacity of the concrete in compression can be efficiently utilized across the entire depth of the concrete sections when all loads act on the structure.

The development of early cracks in reinforced concrete due to non-compatibility in the strains of steel and concrete was perhaps the starting point for the development of a new material like "prestressed concrete".

Prestressed concrete is not a new concept. Dating back to 1872, when P. H. Jackson, an engineer from California, patented a prestressing system that used a tie rod to construct beams or arches from individual blocks. After a long lapse of time during which little progress was made because of the unavailability of high-strength steel to overcome prestress losses, R. E. Dill of Alexandria, Nebraska, recognized the effect of the shrinkage and creep (transverse material flow) of concrete on the loss of prestress. In the early 1920s, W. H. Hewett of Minneapolis developed the principles of circular prestressing.

Eugene Freyssinet proposed methods to overcome prestress losses through the use of high-strength and high-ductility steels in 1926—1928. In 1940, he introduced the new well-known and well-accepted Freyssinet system.

Prestressed concrete is an improved form of reinforcement. Steel rods are bent into the shapes to provide necessary degree of tensile strength. They are then used to prestress concrete, usually by one of two different methods. The first is to leave channels in a concrete beam that correspond to the shapes of the steel rods. When the rods are running through the channels, they are then bonded to the concrete by filling the channels with grout, a thin mortar of binding agent. In the other (and more common) method, the prestressed steel rods are placed in the lower part of a form that corresponds to the shape of the finished structure, and the concrete is poured around them. Two methods are referred to as "pre-tensioned method" and "post-tensioned method". Because prestressed concrete is so economical, it is a highly desirable material.

From the preceding discussion, it is plain that permanent stresses in the prestressed structural member are created before the full dead and live loads are applied in order to eliminate or considerably reduce the net tensile stresses. With reinforced concrete, it is assumed that the tensile strength of the concrete is negligible and disregarded. This is because the tensile forces resulting from the bending moments are resisted by the bond created in the reinforcement process. Cracking and deflection are therefore essentially irrecoverable in reinforced concrete once the member has reached its limit state at service load.

The reinforcement in the reinforced concrete member does not exert any force of its own on the member, contrary to the action of prestressing steel. The steel required to produce the prestressing force in the prestressed member actively preloads the member, permitting a relatively high controlled recovery of cracking and deflection. Once the flexural tensile strength of the concrete is exceeded, the prestressed member starts to act like a reinforced concrete element.

Two types of bond stress must be considered in the case of prestressed concrete.

The first of these is referred to as "transfer bond stress" and has the function of transferring the force in a pre-tensioned tendon to the concrete. The second type of bond is termed "flexural bond stress" and comes into existence in pre-tensioned and bonded, post-tensioned members when the members are subjected to external loads.

Bond stresses also occur between the tendons and the concrete in both pre-tensioned and bonded, post-tensioned members, as a result of changes in the external load. There are of course no transfer bond stresses in post-tensioned members, since the end anchorage device relatively low in prestressed members for loads less than the cracking load, there is an abrupt and significant increase in these bond stresses after the cracking load is exceeded. Because of the indeterminancy which results from the plasticity of the concrete for loads exceeding the cracking load, accurate computation of the flexural-bond stresses can not be made under such conditions. Again, tests must be relied upon as a guide for design.

Prestressed concrete uses less steel and less concrete. Due to the utilization of concrete in the tension zone, a saving of 15 to 30 percent in concrete is possible in comparison with reinforced concrete. The savings in steel are even higher, 60 to 80 percent, mainly due to the high permissible stresses allowed in the high tensile wires. Although there is considerable saving in the quantity of materials used in prestressed concrete members in comparison with reinforced concrete members, the economy in cost is not that significant due to the additional costs incurred for the high strength concrete high tensile steel, anchorages, and other hardware required for the production of prestressed members. In spite of these additional costs, if a large enough number of precast units are manufactured. The difference between at least the initial costs of prestressed and reinforced concrete systems is usually not very large. And the indirect long-term savings are quite substantial, because less maintenance is needed, a longer working life is possible due to better quality control of the concrete, and lighter foundations are achieved due to the smaller cumulative weight of the superstructure.

The economy of prestressed concrete is also well established for long-span structures. According to Dean, standardized precast bridge beams between 10 and

30 meters long and precast prestressed piles have proved to be economical than steel and reinforced concrete in the United States. According to Abeles, precast prestressed concrete is economical for floors, roofs and bridges of spans up to 30 meters and for cast in situ work, it applies to spans up to 100 meters. In the long span range, prestressed concrete is generally economical in comparison with reinforced concrete and steel construction.

Prestressed concrete offers great technical advantages in comparison with other forms of construction, such as reinforced concrete and steel. In the case of fully prestressed members, free from tensile stresses under working loads, the cross-section is more efficiently utilized when compared with a reinforced concrete section which is cracked under working loads. Within certain limits, a permanent dead load may be counteracted by increasing the eccentricity of the prestressing force in a prestressed structural element, thus effecting saving in the use of materials.

A prestressed concrete flexural member is stiffer under working loads than a reinforced concrete member of the same depth. However, after the onset of cracking, the flexural behavior of a prestressed member is similar to that of a reinforced concrete member. Prestressed concrete members possess improved resistance to shearing forces, due to the effect of compressive prestress, which reduces principal tensile stress. The use of curved cables, particularly in long-span members helps to reduce the shear forces developed at the support sections.

The use of high strength concrete and steel in prestressed members results in lighter and slender members than could be possible by using reinforced concrete. The two structural features of prestressed concrete, namely high strength concrete and freedom from cracks, contributes to the improved durability of the structure under aggressive environmental conditions. Prestressing of concrete improves the ability of the material for energy absorption under impact loads. The ability to resist repeated working loads has been proved to be as good in prestressed as in reinforced concrete.

Prestressed concrete has made it possible to develop buildings with unusual shapes, like some of the modern sports arenas, with large spaces unbroken by any obstructing supports. The uses for this relatively new structural method are constantly being developed.

Today, prestressed concrete is used in buildings, underground structures, TV towers, floating storage and offshore structures, power stations, nuclear reactor vessels, and numerous types of bridge systems including segmental and cable-stayed bridges. They demonstrate the versatility of the prestressing concept and its all-encompassing application. The success in the development and construction of all these structures has been due in no small measures to the advances in the technology of materials, particularly prestressing steel, and the accumulated knowledge in estimating the short-term and long-term losses in the prestressing forces.

课文选自：余家欢. 土木工程专业英语 [M]. 北京：清华大学出版社，2017.

阅读选自：宿晓萍，赵庆明. 土木工程专业英语 [M]. 北京：北京大学出版社，2017.

Lesson 4 :
Steel Structures

Steel structure refers to a building in which steel plays the leading role. The early development of high-rise buildings began with structural steel framing. With the development of science and technology, there are more and more types of steel structure. Steel frame building and space structure are within it.

Steel frame building consists of a skeletal frame work which carries all loads to which the building is subjected. It is made up of separate elements-beams, columns, portals, trusses, plates, bracing, purlins, etc. Beams-members carrying lateral loads in bending and shear columns-members carrying axial loads in compressing and bending, portals and trusses-members carrying lateral loads, plates-members supporting wall, bracing, together with columns and trusses, resist wind loads and stabilize the building, purlins-members carrying roof sheet. These elements must be joined together so as to be in position and carry loads without bulking out of the plane. Steel frame structures are extensively used in office, flat, industry, hospital, etc.

One of the visible changes on steel structure is the remarkable trend towards greater use of space structures, this trend is growing as a result of architectural preference. This is partly due, no doubt, to reaction from beam-column systems of previous decades, but also due to the realization of the advantages of spaces structures. Structural engineers realized many years ago the fact that space structure requires less material than the conventional linear systems and that, if properly designed, prefabricated space structures can be highly economical in cost.

There are various types of space structures, differing in their behavior under load and requiring different methods of analysis. Double-layer grids are typical examples of space structures and also one of the most popular forms of space frames. They are frequently used nowadays all over the world for covering large-span industrial buildings, sports halls, churches and exhibition centers.

Present experience shows that in many countries double-layer grids structures can complete very successfully on a cost basis with more conventional systems, providing at the same time additional advantages, such as greater rigidity, simplification of erection and the possibility of covering larger spans.

Design of connection is important in steel structure design. Joints are designed to transmit axialload, shear, moment and torsion as the frame analysis and design of members. In general, a pinned joint transmits axial load and shear; a rigid joint transmits all actions. Sliding joints are often required where provision for expansion is needed. Joints are made by rivets, bolts and welds.

Welding results in important advantages, for example, the structure is cleaner and better looking, maintenance costs are lower. But the defects which occur in welds lead to a reduction in strength of the joint and they may also initiate failure due to brittle fracture of fatigue. The main defects include: slag inclusions, gas pockets, incomplete penetration, under cutting, residual stress and distortion.

The two types of bolts in general use in structural steelwork are black bolts and high-strength fraction grip bolts. The black bolts are forged from round bars with machined threads on bolts and nuts, they are used in holes with 2 mm clearance, they may be used in shear, tension, torsion or in combined shear, tension and torsion. The high-strength fraction grip bolts must be tightened to provide the necessary shank tension, the bolts are used in holes less than 2 mm clearance and in shear, tension, bending or in combined shear, tension and bending. Joints with high-strength fraction grip bolts provide higher capacity and little deformation compared with black bolts.

Steel sections include hot rolled and formed sections, cold-rolled sections and build-up sections. The hot rolled and formed sections, such as, equal and unequal angles, channels, structural tees, circular, square and rectangular hollow sections. The cold-rolled sections, such as zed, lipped channel. Build-up column and box column are build-up sections. For asymmetrical sections, the neutral axis must be located first. For build-up sections, the properties must be calculated.

1. Phrases and Expressions

bracing |breɪsɪŋ| *n.* 支撑
purlin |pɜːlɪn| *n.* 檩条
truss |trʌs| *n.* 桁架
lateral load *n.* 横向荷载
prefabricated space structures *n.* 预制空间结构
double-layer grid *n.* 双层网格
rigidity |rɪˈdʒɪdətɪ| *n.* 刚度
erection |ɪˈrekʃn| *n.* 安装
shearload *n.* 剪力
torsion |ˈtɔːrʃn| *n.* 扭矩
rigid joint *n.* 刚性节点
pinned joint *n.* 铰接节点
fatigue |fəˈtiːg| *n.* 疲劳
slag inclusion *n.* 夹渣
gas pocket *n.* 气孔
residual stress *n.* 残余应力
thread |θred| *n.* 螺纹
shank |ʃæŋk| *n.* 螺杆
deformation |diːfɔːrˈmeɪʃn| *n.* 变形

2. Please translate the following sentences into Chinese

(1) The early development of high-rise buildings began with structural steel framing.

(2) One of the visible changes on steel structure is the remarkable trend towards greater use of space structures.

(3) There are various types of space structures, differing in their behavior under load and requiring different methods of analysis.

(4) Welding results in important advantages, for example, the structure is cleaner and better looking, maintenance costs are lower.

(5) Joints are designed to transmit axial load, shear, moment and torsion as the frame analysis and design of members.

3. Please translate the following sentences into English

（1）柱子对局部屈曲的敏感性可通过其横截面的宽厚比来衡量。

（2）显然，与梁或受拉构件相比，柱子在结构中是较为关键的构件。

（3）轴向受压柱有三种常见的失效模式：弯曲屈曲、局部屈曲和扭转屈曲。

（4）W 型钢具有最经济的截面形式，因此它已经大量地替代了槽钢和 S 型钢。

（5）这种构件几乎没有刚度，在其自重作用下容易产生变形，影响了结构的外观。

4. Extension

tie rod *n.* 拉杆，系杆
sleeve nut *n.* 套筒螺帽
rolled shape *n.* 轧制钢材
slenderness ratio *n.* 长细比
buckling |bʌklɪŋ| *n.* 屈曲
stability |stəˈbɪlətɪ| *n.* 稳定性
solid-web steel column *n.* 实腹式钢柱
splice plate *n.* 拼接板
wrench |rentʃ| *n.* 扳手
gusset |ˈgʌsɪt| *n.* 节点板
end connection *n.* 端部连接
diagonal bracing *n.* 斜撑
automatic arc welding *n.* 自动电弧焊
non-destructive evaluation *n.* 无损检测
manual arc welding *n.* 手工电弧焊

5. Reading material

High Strength Bolts and Welding

The joints obtained using high-strength bolts are superior to riveted joints in performance and economy and they are the leading field method of fastening structural steel members. C. Batho and E. H. Bateman first claimed in 1934 that high-strength bolts could satisfactorily be used for the assembly of steel structures, but it was not until 1947 that the Research Council on Riveted and Bolted Structural Joints of the Engineering Foundation was established. This group issued their first specifications in 1951, and high-strength bolts were adopted by both building and bridge engineers for both static and dynamic loadings with amazing speed. They not only quickly became the leading method of making field connections, but also they were found to have many applications for shop connections. The construction of the Mackinac Bridge in Michigan involved the use of more than one million high-strength bolts.

Connections that were formerly made with ordinary bolts and nuts were not satisfying when they were subjected to vibratory loads because the nuts frequently became loose. For many years this problem was dealt with by using some types of locknut, but the modern high-strength bolts furnish a far superior solution.

Among the many advantages of high-strength bolts, partly explaining their great success, are the following.

(1) Smaller crews are involved as compared with riveting. Two two-person bolting crews can easily turn out over twice as many bolts in a day as the number of rivets driven by the standard four-person riveting crew. The result is quicker steel erection.

(2) In comparison with rivets, fewer bolts are needed to provide the same strength.

(3) Good bolted joints can be made by people with a great deal less training and experience than is necessary to produce welded and riveted connections of equal quality. The proper installation of high-strength bolts can be learned in a matter of hours.

(4) No erection bolts are required that may have to be later removed (depending on specifications) as in welded joints.

(5) Though quite noisy, bolting is not nearly as bad as riveting.

(6) Cheaper equipment is used to make bolted connections.

(7) No fire hazard is present, nor danger from the tossing of hot rivets.

(8) Tests on riveted joints and fully tensioned bolted joints under identical conditions definitely show that bolted joints have a higher fatigue strength. Their fatigue strength is also equal to or greater than that obtained with equivalent welded joints.

(9) Where structures are to be later altered or disassembled, changes in connections are quite simple because of the ease of bolt removal.

The American Welding Society' s Structural Welding Code is the generally recognized standard for welding in the United States. The LRFD Specification clearly states that the provisions of the AWS Code apply under the LRFD Specification with a very few minor exceptions, and these are listed in LRFD Specification J2. Both the AWS and the AASHTO Specifications cover dynamically loaded structures. Normally, however, the AWS Specification is used for designing buildings subject to dynamic loads unless the contract documents state otherwise.

Although both gas and arc welding are available, almost all structural welding is arc welding. Sir Humphry Davy discovered in 1801 how to create an electric arc by bringing close together two terminals of an electric circuitof relatively high voltage. Although he is generally given credit for the development of modern welding, a good many years elapsed after his discovery before welding was actually performed with the electric arc (His work was of the greatest importance to the modern structural world, but it is interesting to note that many people say his greatest discovery was not the electric arc, but rather a laboratory assistant whose name was Michael Faraday) .

Several Europeans formed welds of one type or another in the 1880s with the electric arc, while in the United States the first patent for arc welding was given to Charles Coffin of Detroit in 1889.

In electric-arc welding, the metallic rod, which is used as the electrode, melts off into the joint as it is being made. When gas welding is used it is necessary to introduce a metal rod known as a filler or welding rod.

In gas welding a mixture of oxygen and some suitable type of gas is burned at the tip of a torch or blowpipe held in the welder's hand or by an automatic machine. The gas used in structural welding is probably acetylene, and the process is called oxyacetylene welding. The flame produced can be used for flame cutting of metals as well as for welding. Gas welding is rather easy to learn and the equipment used is rather inexpensive. It is, however, a somewhat slow process as compared with other means of welding, and normally is used for repair and maintenance work and not for the fabrication and erection of large steel structures.

In arc welding, an electric arc is formed between the pieces being welded and an electrode held in the operator's hand with some type of holder, or by an automatic machine. The arc is a continuous spark that upon contact brings the electrode and the pieces being welded to the melting point. The resistance of the air or gas between the electrode and the pieces being welded changes the electrical energy into heat. At the end of the electrode melts, small droplets or globules of the molten metal are formed and are actually forced by the arc across to the pieces being connected, penetrating the molten metal to become a part of the weld. The amount of penetration can be controlled by the amount of current consumed. Since the molten droplets of the electrodes are actually propelled to the weld, arc welding can be successfully used for overhead work.

A pool of molten steel can hold a fairly large amount of gases in solution, and if not protected from the surrounding air will chemically combine with oxygen and nitrogen. After cooling the welds will be relatively porous due to the little pockets formed by the gases. Such welds are relatively brittle and have much less resistance to corrosion. A weld can be shielded by using an electrode coated with certain mineral compounds. The electric arc causes the coating to melt and creates an inert gas or vapor around the area being welded. The vapor acts as a shield around the molten metal and keeps it from coming freely in contact with the surrounding air. It also deposits a slag in the molten metal, which has less density than the base metal and comes to the surface to protect the weld from the air while the weld cools. After cooling, the slag can easily be removed by peening and wire brushing (such removal being absolutely necessary before painting or application of another weld layer) . Shielded metal arc welding is frequently abbreviated here with the letters SMAW.

The type of welding electrode used is very important as it decidedly affects the weld properties such as strength, ductility, and corrosion resistance. Quite a number of different types of electrodes are manufactured, the type to be used for a certain job being dependent upon the type of metal being welded, the amount of material that needs to be added, the position of the work, etc.

The heavily coated electrodes are normally used in structural welding because the melting of their coatings produces very satisfactory vapor shields around the work as well as slag in the weld. The resulting welds are stronger, more resistant to corrosion, and more ductile than those produced with lightly coated electrodes. When the lightly coated electrodes are used, no attempt is made to prevent oxidation and no slag is formed. The electrodes are lightly coated with some arc-stabilizing chemical such as lime.

Submerged (or hidden) arc welding (SAW) is an automatic process in which the arc is covered with a mound of granular fusible material and thus hidden from view. A bare metal electrode is fed from a reel and melted and deposited as filler material. The electrode, power source, and a hopper of flux are attached to a frame that is placed on rollers and that moves at a certain rate as the weld is formed. SAW welds are quickly and efficiently made and are of high quality, exhibiting high impact strength and corrosion resistance and good ductility. Furthermore, they provide deeper penetration with the result that the area effective in resisting loads is larger. A large percentage of the welding done for bridge structures is SAW. If a single electrode is used, the size of the weld obtained with a single pass is limited. Multiple electrodes may be used, however, permitting much larger welds.

Welds made by the SAW process (automatic or semiautomatic) are consistently of high quality and are very suitable for long welds. One disadvantage is that the work must be positioned for near flat or horizontal welding.

Another type of welding is flux-cored arc welding (FCAW) . In this process a flux-filled steel tube electrode is continuously fed from a reel. Gas shielding and slag are formed from the flux. The AWS Specification provides limiting sizes for welding electrode diameters and weld sizes, as well as other requirements pertaining to welding procedures.

课文选自：霍俊芳，姜丽云. 土木工程专业英语 [M]. 北京：北京大学出版社，2010.

阅读选自：苏小卒. 土木工程专业英语 [M]. 上海：同济大学出版社，2000.

Lesson 5 :
Bridge Structures

Birth of Bridges

The first man-made bridge was probably a tree trunk or flat stone laid across a stream. No doubt it was made many thousands of years before the birth of Christ. Even before that, primitive man must have wondered at natural arches such as the Pont d' Arc at Ardeche that has a span of 194 feet rising 111 feet over the river. But ages would have passed before some pioneer jammed two stones together like an inverted "V" across a narrow brook and so built the first arch bridge.

According to Degrand, the earliest bridge on record is that built on the Nile by Menes, the first King of the Egyptians, about 2650 B. C., but no details are forthcoming. A detailed description of another bridge built about five centuries later is given by Diodorus Siculus — the fabulous bridge built by Semiramis, Queen of Babylon, over the Euphrates. Herodotus ascribes this bridge to Queen Nitocris who ruled five generations later. First the river was diverted well above the city into an artificial lake, so that the piers of the bridge could be built in the river bed. The stones of the piers were bonded together by iron bars soldered in with lead. The deck was of timber, cedar, cypress and palm, and was no less than 30 feet wide. Part of it was removable and was taken up each night to afford protection from robbers. When the bridge was finished the river was brought back into its original channel. So the record reads to-day. How much is true and how much due to embellishment through the ages we shall never know, but there is no doubt that a remarkable bridge was built 4,000 years ago in Babylon.

Bridge Types

We can only speculate on these beginnings. We see primitive suspension bridges made of twisted creepers or lianas tied to tree trunks on either side of a gorge and spanning it like a perilously hung cobweb. But when these bridges were evolved or which came first we cannot say with certainty. We only know that the three types, beam or girder bridges (typified by a tree trunk across a stream), arch bridges, and suspension bridges, have been known and built from the earliest times of which we have any record. In their simplest form beam or girder bridges are called simple spans; if two or more are joined together over the piers they become continuous; or they may be built to form cantilever bridges. These, however, are only varieties of girder bridges and do not constitute a different type. The three types, girder, arch and suspension, may be varied and combined to assist each other in the same structure, and down the years materials of construction have evolved from those ready to hand, such as timber and stone, to manufactured materials such as brick, concrete, iron and steel.

Beam Bridges

A simple single-span bridge may be of steel (probably a plate girder), reinforced concrete or prestressed concrete. In steel the maximum span for a simple beam bridge is usually about 100 ft (although bridges with longer spans have been built) . When, however, the spans are large, a continuous girder is usually adopted. A plate-girder bridge in Germany has a central span of 354 ft and side spans of 295 ft.

For spans between supporting piers above about 150 ft, the truss is often used and the material is invariably steel. A bridge over the Ohio in Illinois, completed in 1917, has a simply supported span of 720 ft.

The principle of the cantilever bridge, with and without a suspended span, is illustrated in fig. 5-1 (a) and (b), although the latter is not common in solid beam or girder construction. The piers having been built, the bridge (anchored at A and B) is built out from each pier, and the middle portion of the bridge, called the suspended span, which is usually in one prefabricated unit, is then placed in position. The bridge therefore consists of two anchored cantilevers supporting a beam “suspended” from the ends of the cantilevers. The maximum bending moments and shear forces occur at C and D, and at these points the bridge is usually of greater depth.

When spans are large, thus requiring a great depth of bridge, cantilever bridges are usually constructed of steel trusses (trussed girders) . It is possible in this way to have spans of up to about 1,800 ft between piers. Fig. 5-1 (c) is an example where the cantilevers meet without a middle suspended span. Although this bridge may look like an arch, it is in fact a double-cantilever trussed beam. It may be noted that in cantilever bridges the greatest depth of truss occurs at the main piers because it is at these points that the greatest stresses occur.

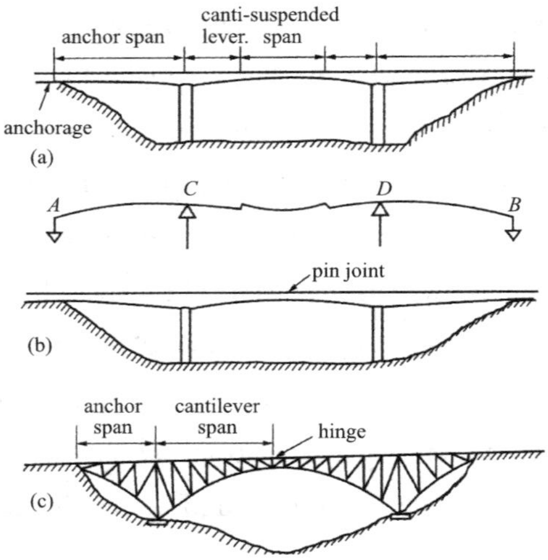

Fig. 5-1 Cantiever bridges

Arch Bridges

In an arch bridge the arch is the main structural member and transmits the loads imposed on it to the abutments at the springing points. The part of the construction above the arch ring where the roadway or railway is at a higher level than the crown of the arch is called the spandrel.

Since steel and reinforced concrete are capable of taking tension, the arch rings can be very much thinner than masonry construction. The braced spandrel bridge is usually constructed in steel, as it is also the bridge where the roadway is supported by hangers from the structural arch.

Another type of arched bridge is the stiffened tiedarch, which is often called a bow-string girder. In an archery bow the string prevents the bow from flattening out. In a similar manner, the road-supporting horizontal girders are made strong enough to absorb the arch thrusts, and therefore the reactions on the piers and abutments are vertical.

Suspension Bridges

When spans are large, about 2,000 ft or more, suspension bridges are the most economical, but they can, of course, be used for smaller spans. Usually, there is a central span with two side spans and the cables passing over the top of the supporting piers are anchored in tunnels or by other means. Since the cables pull on each pier, the load on the pier is almost entirely vertical. The roadway is suspended from the inclined cables by vertical hangers.

All bridge design tasks can be included in four areas: planning, selection of bridge type, selection of materials and analysis of forces.

Bridge Design

When designing and constructing a long-span bridge the great weight of the structure, the dynamic effects of moving loads such as locomotives or motor vehicles, and the aerodynamic effects of wind pressure give rise to problems which call for the greatest knowledge and ingenuity in their solution.

Planning

The construction of long-span bridge is a great achievement, and the history of bridge building includes many human, romantic and even tragic stories. The first step leading to the construction of a modern major bridge is a comprehensive study to determine whether a bridge is needed. If it is to be a highway bridge, in the United States for example, a planning study is initiated by a state bridge authority, possibly in cooperation with local governments or the federal government. Studies are made to estimate the amount of bridge traffic, the relief of jammed traffic in nearby highway networks, the effects on the regional economy, and the cost of the bridge. The means for financing the project, such as public taxes or sale of revenue bonds repaid by toll charges, are considered. If the studies lead to a decision to go ahead with the project, the land needed for the bridge and its approaches is acquired at the selected site. At this point, field engineering work is started. Accurate land surveys are made. Tides, flood conditions, currents and other characteristics of the waterway are carefully studied. Boring samples of soil and rock are taken at possible foundation locations, both on land and under the water.

Selection of Bridge Types

The chief factors in deciding whether a bridge will be built as a girder, cantilever, truss, arch, suspension, or some other types are: (1) location: for example, across a river; (2) purposes: for example, a bridge for carrying motor vehicles; (3) span length;

(4) strength of available materials; (5) cost; (6) beauty and harmony with the location.

Each type of bridge is most effective and economical only within a certain range of span lengths, as shown Table 5-1.

Best span ranges of different bridges **Table 5-1**

Bridge Type	Best Span Range					
	(feet)			(meters)		
Girder	20	to	1,000	6.1	to	304.8
Rigid Frame	80	to	300	24.4	to	91.4
Arch	200	to	1,000	61.0	to	304.8
Truss	200	to	1,400	61.0	to	426.7
Cantilever	500	to	1,800	152.4	to	548.6
Suspension	1,000	to	5,000	304.8	to	1,524.0

As indicated in the table, there is a considerable overlap in the range of applicability of the various types. In some areas, alternative preliminary designs are prepared for several types of bridge in order to have a better basis for making the final selection.

Selection of Materials

The bridge designer can select from a number of modern high-strength materials, including concrete, steel and a wide variety of corrosion resistant alloy steels.

For the Verrazano-Narrows Bridge, for example, the designer used at least seven different kinds of alloy steel, one of which has a yield strength of 50,000 pounds per square inch (psi) (3,515 kgs/sq cm) and does not need to be painted because an oxide coating forms on its surface and inhibits corrosion. The designer also can select steel wires for suspension cables that have tensile strength up to 250,000 psi (17,577 kgs/sq cm) .

Concrete with compressive strengths as high as 8,000 psi (562.5kgs/sq cm) can now be produced for use in bridges, and it can be given high durability against chipping and weathering by the addition of special chemical agents and control of the hardening process. Concrete that has been prestressed and reinforced with steel wires has a much higher tensile strength.

Other useful materials for bridges include aluminum alloys and wood. Modern structural aluminum alloys have yield strengths exceeding 40,000 psi (2,812 kgs/sq cm) . Laminated strips of wood glued together can be made into beams with strengths twice that of natural timbers; glue-laminated southern pine, for example, can bear working stresses approaching 3,000 psi (210.9 kgs/sq cm) .

Analysis of Forces

A bridge must resist a complex combination of tension, compression, bending, shear and torsion forces. In addition, the structure must provide a safety factor as insurance against failure. The calculation of the precise nature of the individual stresses and strains in the structure, called analysis, is perhaps the most technically complex aspect of bridge building. The goal of the analysis is to determine all of the forces that may act on each structural member.

The forces that act on bridge structural members are produced by two kinds of loads — static and dynamic. The static load — the dead weight of the bridge structure itself — is usually the greatest load. The dynamic, or live load, has components, including vehicles carried by the bridge, wind forces, and accumulations of ice and snow.

Although the total weight of the vehicles moving over a bridge at any time is generally a small fraction of the static and dynamic load, it presents special problems to the bridge designer because of the vibration and impact stresses created by moving vehicles. For example, the severe impacts caused by irregularities of vehicle motion or bumps in the roadway may momentarily double the effect of the live load on the bridge.

Wind exerts force on a bridge both directly by striking the bridge structure and indirectly by striking vehicles that are crossing the bridge. If the wind induces aeroelastic vibration, as in the case of the Tacoma Narrows Bridge, its effect may be greatly amplified. Because of this danger, the bridge designer makes provisions for the strongest winds that may occur at the bridge location. Other forces that may act on the bridge, such as stresses created by earthquake tremors, must also be provided for.

Special attention must often be paid to the design of bridge piers, since heavy loads may be imposed on them by currents, waves and floating ice and debris. Occasionally a pier may even be hit by a passing ship.

Electronic computers are playing an ever-increasing role in assisting bridge designers in the analysis of forces. The use of precise model testing, particularly for studying the dynamic behavior of brides, also helps designers. A scaled-down model of bridge is constructed, and various gauges to measure strains, accelerations and deformations are placed on the model. The model bridge is then subjected to various sealed-down loads or dynamic conditions to find out what may happen. Wind tunnel test may also be made to ensure that nothing like the Tacoma Narrows Bridge failure can occur. With modern technological aids, there is much less possibility of bridge failure than in the past.

1. Phrases and Expressions

arch ring　*n.* 拱圈
brace　|breɪs|　*n. v.* 支撑，联结
cantilever bridge　悬臂桥
continuous bridge　连续梁桥
hanger　|ˈhæŋər|　*n.* 吊杆
long span　大跨径
pin joint　铰接结点
plate girder　板梁
rigid frame　刚构（桥），刚架（桥）
side span　边跨
span　|spæn|　*n.* 跨径
suspended span　悬臂跨
suspension　|səˈspenʃn|　悬索（桥），吊桥
tied arch　系杆拱

2. Please translate the following sentences into Chinese

(1) In their simplest form beam or girder bridges are called simple spans. If two or more are joined together over the piers they become continuous; or they may be built to cantilever bridges.

(2) For spans between supporting piers above about 150 ft, the truss is often used and the material is invariably steel.

(3) In an arch bridge the arch is the main structural member and transmits the loads imposed on it to the abutments at the springing points.

(4) All bridge design tasks can be included in four areas: planning, selection of bridge type, selection of materials and analysis of forces.

(5) The calculation of the precise nature of the individual stresses and strains in the structure, called analysis, is perhaps the most technically complex aspects of bridge building.

3. Please translate the following sentences into English

（1）悬臂桁架梁桥的最大桁高位于桥墩处，原因是这里的应力最大。

（2）悬索桥通常由一个中跨、两个边跨组成，其主缆架在桥塔上，两端固定于锚碇或桥身上。

（3）基础的设计和施工对于桥梁至关重要，特别是当基础位于河流或海域时。

（4）现代高强建筑材料如混凝土、钢材和多种多样的耐腐蚀合金钢均可被桥梁设计师选用于造桥。

（5）尽管从静力上讲交通荷载重量在桥梁总荷载中占比例不大，但车辆行驶产生的动力效应不容忽视。

4. Extension

box girder　箱梁

cable-stayed bridge　斜拉桥

carriageway |ˈkærɪdʒweɪ| *n.* 行车道
clearance heigh 净空高度
diaphragm |ˈdaɪəfræm| *n.* 横隔板
drainage system 排水系统
expansion joint 伸缩缝
high-rise pier 高墩
hollow pier 空心桥墩
orthotropic deck 正交异性板
pier coping 桥墩盖梁
pile cap 桩承台
rise-span ratio 矢跨比
sidewalk |ˈsaɪdwɔːk| *n.* 人行道
stiffening girder *n.* 加劲梁
substructure |ˈsʌbstrʌktʃər| *n.* 下部结构
superstructure |ˈsuːpərstrʌktʃər| *n.* 上部结构
under clearance 桥下净空
viaduct |ˈvaɪədʌkt| *n.* 高架桥
web |web| *n.* 腹板

5. Reading material

Birth of the Modern Bridge

The Renaissance — that great upsurge of interest in art and science and sudden zest to live life to its full — had its origin in Florence in the fifteenth century. It was born out of the vision of men in the tradition of Leonardo da Vinci: manifested in the paintings and sculpture of Michelangelo and Donatello, proclaimed in the writings of Dante, and reached its architectural perfection in the works of Palladio and Vignola. Amongst bridges it was responsible for the finest achievements of the sixteenth century — which means it is the sturdy Pant Notre Dame and Pont Neuf in Paris, the lovely Santa Trinita bridge in Florence itself and the picturesque Rialto bridge spanning the Grand Canal in Venice.

Following on this re-awakening of spirit and energy, bridge building in the seventeenth and eighteenth centuries at last became a science. Lines of thrust were plotted, the strength of materials and foundations closely estimated. In 1716 was formed the Corps des Ingénieurs des Ponts et Chaussées, the members of which were engineers trained in Paris, to whom plans of all roads, bridges and canals in central France had to be submitted for approval. Some years later they founded the first engineering school in the world, to which Jean Perronet was appointed Director. By the middle of the eighteenth century, bridge building in masonry reached its zenith in such masterpieces as Perronet' s bridge at Neuilly and William Edwards' s at Pontypridd, By this time, as we shall see, bridge trusses of timber, first evolved by Palladio, had been developed by the brothers Grubenmann in Switzerland. Their use continued, particularly in North America, until late in the nineteenth century. But long before that the first iron bridge had been built in England and a new age had begun.

Let us first see the effect of the Renaissance on the bridges in France. In the days when Paris, or Lutetia as it was then called, was the headquarters of the Roman occupation, it was confined to the island in the Seine now known as La Cité. Under Julius Caesar. and even up to the fourth century A. D., there was only one river crossing, this was provided by two timber bridges, which linked the island to the river banks. Ten centuries passed before the increase in traffic led to the building of the first Pont Notre Dame, so called because it gave direct access to the cathedral. Sixty houses were built on the bridge, which was of timber, and Charles Ⅵ conceded the rents from them to the city authorities for its maintenance. In spite of many cajolings and warnings, however, no repairs were made, and in 1499 the whole structure collapsed into the river. By this time the Renaissance had spread to Paris, and under its stimulus the decision was made to rebuild the bridge in stone. The requisite funds were raised by means of a toll on cattle, fish and salt.

The work was entrusted to Fra Giovanni Giocondo, who on account of his fame in both science and the arts, had been invited to Paris to advise Charles Ⅷ on the Italian style in building. Giocondo had restored the old Roman bridge at Verona and was subsequently engaged in the strengthening of the foundations of St Peter' s in Rome.

The Pont Notre Dame consisted of six arches, the central four having spans of 57 feet. Discussion arose as to whether the foundations should be excavated to bed rock or whether they should be piled. This was one of numerous disputes between Giocondo and Didier de Felin, one of the French superintendents, in which Giocondo usually seemed to get his way. Piles consisting of heavy tree trunks were finally adopted and they were driven inside cofferdams from which the water had been expelled by means of horse-operated pumps. The masonry piers were then built in the dry on footings of rubble concrete.

The bridge was completed in 1507, after seven years' work, and two fine rows of our-story houses with cellars were built lining the roadway. Although the backs of the houses overhung the river, they reduced the width of highway available for traffic from 75 feet to 25 feet, which was evidently considered sufficient at that time. It was not until 1786 that the volume of traffic brought about the demolition of the houses, and the roadway was rebuilt 42 feet wide with footways and parapets on either side. In spite of the designer's efforts to avoid scour around the piers by leaving as wide a waterway as possible, the city authorities installed a weir beneath one arch and mills under two others. When the old arches were demolished in 1853, however, their foundations were found to be in such good condition, after 350 years, that they were built into the new bridge that we see to-day.

The second great Renaissance bridge in Paris is the famous Pont Neuf built in 1578—1604 to the order of Henry Ⅲ at the downstream end of the Ile de la Cité. The north and south parts of the bridge consist of seven and five spans respectively. The arches are all built on a slight skew, and vary in span from 32 to 64 feet, the piers also vary in thickness. The roadway was 66 feet wide with footways raised two steps above it on either side.

The bridge was designed by Jacques Androuet du Cerceau but the commissioners appointed a board of experts, consisting of master carpenters, master masons and others to advise on the design and execution of the work. One of the first things to decide was the type of foundation. The three methods then in use were to place the ashlar foundation stones (a) directly on the subsoil, (b) on a timber grillage, (c) on timber piles. In spite of the fact that an extra depth of only ten feet would have reached the underlying rock, the board decided to lay the masonry on grillages of timber. Only one of them, Lescot, appears to have had doubts about the quality of the subsoil and would have preferred to use piles. The piers were built inside timber cofferdams, which were driven to refusal by means of a mechanical pile driver. After pumping the cofferdams out and excavating the soft subsoil, a timber grillage was laid down, on top of which the masonry was built. In spite of the expert' s advice, in a few years two of the piers were so badly undermined by scour that they had to be reconstructed even before the bridge was finished. To do this another cofferdam was built around the upstream end of the piers, inside which the original work was repaired. Permanent sheet piling was then driven close against the walls of the piers and anchored to them by iron ties to protect the base against scour. The absence of bearing piles below the foundations proved a permanent weakness, however, and repairs had to be carried out due to settlement on several subsequent occasions.

The building of the superstructure presented no unusual problems, although it was found many years later that parts of the work had been scamped, and that some of the arches were less than the specified thickness. The bridge was opened in 1605 and at once established itself as the main artery of the city. For more than 200 years it was one of the busiest and liveliest centers of Paris. Between the rows of little shops flanking the roadway flowed a ceaseless stream of the city traffic-carts, horsemen, stalls, were thronged with people of every class and profession. Priests and shopkeepers rubbed shoulders with lackeys, artisans and sailors. Amongst the honest citizens of Paris mingled loafers, cloak-snatchers and pickpockets awaiting their opportunity. So serious did the nuisance become that in 1640 an Act was passed forbidding "lackeys, soldiers, vagabonds and all others playing cards, dice, or other forbidden games on the Pont Neuf". During the Revolution, carts rumbled continually across the bridge carrying victims to the guillotine.

In 1848 an extensive reconstruction of the whole bridge was carried out. Flatter elliptical arches were built in place of the original circular ones on the north side, in order to reduce the rise and hence the gradient for traffic. The roadway was remodeled, the round shoulders on the pier ends were rebuilt all at the same height above the water, and ornamental parapets and lighting standards added.

At the time of the Renaissance one of the four medieval bridges over the Arno in Florence was the picturesque Ponte Vecchio (1367), lined with its famous goldsmith's shops. This bridge still stands today, little altered since the time of the Medici. Another was the Santa Triniata bridge which Cosimo I, the Grand Duke, ordered his engineer, Bartolomeo Ammanati, to replace. Ammanati had studied under Sansovino, who made a design for the Rialto bridge in Venice and had completed the building of the Pitti Palace in Florence. The river was 320 feet wide at the site and Ammanati designed a most beautiful bridge of three arches, with spans of 87 feet, 96 feet and 86 feet, respectively.

Possibly influenced by the Ponte Vecchio, which had exceptionally low arches for a medieval bridge, Ammanati designed those of Santa Trinita with a rise: span ratio of only 1 to 7 instead of the usual ratio of 1 to 4. He thus avoided a hog-backed road with its steeply inclined approaches, difficult for traffic to negotiate. Furthermore, in order to avoid the apparent weakness of a long flat soffit, he had the courage to adopt the pointed arch of the East, and introduce it into the very home and citadel of the Roman circular arch. By two more strokes of genius he concealed the angel at the crown with carved pendants, and closely integrated arches and piers by starting the curves of the arches vertically from the springings. In effect Ammanati evolved the first "basket-handled" arch — a shape that gives so many advantages that it has since been widely used. It was too much, of course, to expect popular opinion to accept such changes without suspicion. In fact it is said that not until after Napoleon had ordered his heavy artillery across the bridge, during his invasion of Italy, did other vehicles dare to use it!

The foundations were built inside cofferdams of an unusually extensive kind. Two concrete walls, 7 feet thick and about 90 feet apart, were built inside sheet piling right across the river, which was presumably diverted first to one side and then to the other. Cross walls were then built between them, thus forming rectangular cofferdams, which could be pumped out in turn, around the site of each pier. The ground inside was excavated to a depth of 13 feet and piled. The foundation stones of the piers, each measuring 6 feet × 18 inches in plan, were placed on the bearing piles and the masonry brought up to its full height. The arches were then built on seven temporary timber centers spaced 10 feet apart.

Ruskin described the famous Rialto bridge in Venice as "the best building raised in the time of the Grotesque Renaissance; very noble in its simplicity, in its proportions and in its masonry". Its low circular arch crosses the Grand Canal in a span of 88 feet, causing scarcely any interference with the waterway. The rise of the arch is only 21 feet, which allowed sufficient headroom for the state barges and galleys that used the canal when the bridge was built, and is ample today for the gondolas and motor launches that ply beneath it. The bridge is 75 feet wide and carries a central roadway lined with shops and sidewalks. There is, of course, no wheeled traffic in Venice, and wide shallow steps carry the roadway up over the rise of the arch.

Since the twelfth century there had been a succession of bridges on the site, these were of various kinds, including pontoon and timber bridges and an ingenious bridge with a central opening span that had two arms which could be raised like drawbridges to let shipping pass. After a great fire in 1512, which ravaged the Rialto district and threatened the bridge, Giocondo, who had returned to live in Venice, suggested that a stone bridge should be built with shops, as on the Pont Notre Dame.

In the following years a number of eminent men, including Michelangelo, Palladio, Scamozzi and da Ponte prepared designs, but of those that have survived, only that of Antonio da Ponte, which was accepted in 1588, was suited to the site conditions and satisfactorily solved the difficult problem of the foundations. For Venice is a city of many islands formed in soft alluvial ground, not at all an easy subsoil in which to resist the thrust of a low arch, surcharged with the weight of a street of a street of shops. Da Ponte solved the problem by means of extensive piling. He drove as many as 6,000 piles of birch-alder beneath each abutment. The piles were 6 inches or so in diameter, 11 feet long, and driven to refusal. Owing to the close proximity of buildings with shallow foundations. he did not dare excavate to any great depth at the back of the abutments, but formed the ground into steps at three levels, the deepest of which was 16 feet at the canal front. These excavations were made inside cofferdams which were kept reasonably dry by means of large pumps. In order to get 6,000 piles in the base of each abutment they must have been driven so close that they were practically touching each other. This would not be considered good practice today, but it appears to have been effective.

After driving, the pile heads were cut level and capped with three layers of timbers secured by iron clamps. Da Ponte then had wedge-shaped stones with brick backing laid on the timbers, in an endeavour to make the bed joints in the abutment normal to the line of thrust of the arch. This was an outstanding innovation that has frequently been adopted since da Ponte's day. In the great concrete skewbacks beneath the bearings of Sydney Harbour bridge, for instance, the concrete was so placed that every layer presented a number of facets approximately normal to the line of thrust of the arch.

Possibly owing to the envy of Scamozzi, whose design for a bridge on a widespread "floating" foundation had been passed over in favour of da Ponte's piled abutments, construction was not allowed to proceed unhindered. Work was halted while a lengthy enquiry was held into the correctness and adequacy of the project. A great number of witnesses were heard, including da Ponte himself, who presented an admirably clear and logical argument in support of his design. In the outcome, the commission accepted all his proposals and the work was completed in accordance with his plans. A few years after it was opened the bridge was shaken by an earthquake that rocked the city. Shopkeepers fled in terror, only to return after the rumbling and quaking of the ground had ceased, to find the bridge intact and with no sign of injury.

So it stands to this day, with its busy markets at either end, still the main crossing of the canal and daily thronged with the life and bustle of the city. Venice owes much besides the Rialto bridge to Antonio da Ponte. On two occasions in his capacity as curator of public works he saved the Doges Palace from serious damage by fire, fearlessly entering the blazing building and directing the firefighters. In 1589 he rebuilt the prison, and is responsible for the poignant Bridge of Sighs, over which prisoners were taken for trial in the Palace.

The Pont Royal built in Paris towards the end of the seventeenth century may be regarded as the first modern masonry bridge. It was built solely as a bridge, for the purpose of providing a wide roadway for traffic, with as little obstruction as possible of the river. No houses were built on it, the road and footways were kept clear and unencumbered.

The design was made by Jules-Hardouin Mansard and the construction directed by Jacques (IV) Gabriel, Louis XIV' s architect, who had played a large part in the building of the Palace of Versailles. When Gabriel died several months after the commencement of the work, the responsibility for its completion devolved on his widow, Marie de Lisle. Fortunately her brother Pierre, another architect to the king was able to take over and direct the work on her behalf.

The bridge has five "basket-handled" arches varying in span from 64 ft to 72 ft, springing from piers about 14 feet thick. The roadway is 55 feet wide inclusive of footways and parapets. Before work started, a comprehensive specification was drawn up stating exactly how the bridge was to be built, the quality of materials to be used and the estimated cost, which was to be borne by the king. The use of cofferdams was specified for the construction of the foundations: their walls were to be made 9 feet thick, and to consist of clay puddle between a double sheathing of timbers. The cofferdams were to be pumped out and the ground excavated to a depth of 15 feet below low water. Timber bearing piles 10 inches to 12 inches in diameter were to be driven at 18-inch centres over the area of the base of each pier. The pile heads were to be cut level and capped with a timber platform on which the masonry of the pier was built. The kind of stone to be used in every part of the work was specified.

"The builders will use: the hard stone of Saint-Cloud below low water: the hard stone of Bagneux for the piers up to the springings of the arches and for the spandrels, coping, parapets and kerbs; Vergel stone for the body of the arches; Vaugirard rubble-stones for filling the piers and abutments."

The result of collaboration between Mansards, a skilled architect, and Friar Romain, who was general inspector of the works, the bridge was long in advance of its time and served as a model, particularly in its foundations, for many years.

The next advance was made by Jean Perronet (1703—1794) who had been impressed by the proportions of bridges in China, in which the roadway was carried by huge stone slabs on vertical piers. This gave Perronet the idea of building very flat arches, supported on piers that were simply designed to carry their weight, the whole of the horizontal thrust being transmitted through the arches to the abutments at each end. Incidentally anchorage at the ends is one of the method adopted in the design of earthquake-proof bridges today. It enabled Perronet to reduce the thickness of the river piers and so leave a wider waterway for traffic and lessen the risk of scour. At the same time, it opened the way for flatter arches with comparatively high springings. It was essential, however, that the spans should be almost equal, so as to avoid unbalanced the thrust on the piers, and that all the arches should be built before the temporary centering in anyone of them was removed, Perronet was confirmed in these ideas by his experience on the bridge at Mantes, designed by M. Hupeaux, which he completed in 1765. Here the ratio of thickness of piers to span of arch was 1 : 5, but when one of the side spans was almost finished and the central span only just begun, the unbalance thrust pushed the intermediate pier over 5 inches sideways. Perronet decided that if it was necessary to make the piers thicker than this, to enable each arch to be self-supporting, the obstruction to the waterway would be so great that it was much better to make the arches rely on one another and to carry the horizontal thrust through to the abutment. In his bridges, therefore, he reduced the thickness of the piers to as little as 1/10 of the span. Subsequently, however, he advised the Government that abutment piers should be provided at distances of three to four arches apart in long bridges. This would prevent all the other arches from collapsing following the destruction of one of them. It is of interest to note that Jacques (V) Gabriel had designed two abutment piers of this kind in his bridge of eleven arches over the Loire at Blois built in 1716—1724, but the idea was never generally adopted.

In the Neuilly bridge over the Seine, Perronet reduced the thickness of the piers to less than 1/9 of the span of the arches. This became a matter of hot debate at the Assemblee des Ingenieures des Ponts et Chaussées, but unfortunately there is no record of the proceedings. Even after the bridge had been finished, a certain engineer, M. Defer, expressed such concern for its safety that he asked for sentinels to be placed at the ends of the bridge to slow down the speed of carriages and for the roadway to be covered with 4-inch thick matting to damp out the impact of the wheels on the paving stones!

The striking of the arch centres was an event probably unique in bridge building: "The King having expressed a wish to witness the final operations of decentering, a site was prepared where there were set up a tent for His Majesty, one for the princes, one for the ambassadors, one for the lords of the court and the ministers, and others for the public. It was arranged that the centering of the arches should fall consecutively, within a few minutes. Two winches with ropes were set up to the right of each arch and nine men put to each winch. In 3.5 minutes all the staging had fallen away. When it fell, the mass of timber caused the spray to fly as high as the top of the bridge. His Majesty expressed his satisfaction and on his way back to Marly drove over the bridge in his carriage."

Dinner was provided for the spectators and a medal was struck to commemorate the event. Shaw Sparrow points out, however, that the centering was removed much too soon - only eighteen days after the keystones of the arches had been placed and before the mortar had sufficiently hardened. To this he ascribes the settlement of 23 inches that took place at the crown of one of the arches and considers that Perronet' s reputation was saved "not by his good design, nor by his mathematical calculations, but by a rare stroke of good luck!" Perhaps such strokes of luck are not so rare.

Many judges consider that the Pont Sainte-Maxence over the Oise is Perronet' s finest work. Here the arches are flatter, the piers even more slender, and the waterway less obstructed than at Neuilly. Perronet further reduced the volume of the piers by omitting the stone work in the centre, so that the arches appear to spring from pairs of coupled columns. Perronet' s last work was the Pont de la Concorde (1787—1791). Unfortunately the authorities insisted on the piers being made solid, and the rise of the arches being increased by three feet to give more headroom for shipping. These modifications must have been galling to an engineer of such eminence, and they prevent us from seeing the bridge as Perronet designed it. Nevertheless, it is a beautiful bridge. A critic forty years later referred to the central arch as "the most daring construction that has yet been executed". Perronet watched the bridge being built from a little pavilion at one end, here he saw the first stone set in place and here he died after the work had been finished.

In London the need was felt for another bridge over the Thames to ease the congestion on Old London bridge. Its construction was opposed for years, but finally Parliament decided to build a public lottery! The work was entrusted to a young Swiss engineer, Labelye, and began in 1738. A novel method was used for building the foundations. This consisted of floating out a huge timber box, or "chest", measuring 80 feet by 30 feet overall in plan. The base of the box was a solid timber grillage and the sides were detachable and sufficiently high to stand above water level when the chest was sunk. By opening a sluice gate at the bottom, the chest could be flooded and sunk; by closing the sluice valve and pumping out the water it could be floated up again, if desired. After excavating the site of the pier about 6 feet deep, the chest was sunk in position with several courses of masonry built inside it. The sides of the chest were then removed and used for the next pier, the base remaining permanently in the foundation. This idea might well have been successful if the subsoil had been strong enough to stand without piling, and if precautions had been taken to prevent the foundation from being undermined by scour. At Westminster no bearing piles were driven nor any other precaution taken, and serious settlement of the piers resulted. It is perhaps surprising therefore that the bridge survived for a hundred years.

It was not until the turn of the century that John Rennie (1761—1821) upon whom Perronet' s mantle fell, built the Waterloo, Southwark, and new London bridges. There was nothing new in the foundations of these bridges, which consisted of timber rafts on bearing piles. As regards the superstructure, old Waterloo bridge was much the most attractive, as witness the outcry that arose some years ago when the L. C. C. announced their intention of pulling it down. Both Waterloo and the new London bridge consisted of multiple masonry arches, the biggest spans in each being 119 feet and 152 feet respectively. Old Southwark bridge, commenced in 1815, consisted of three cast-iron arches, the central one having a span of 240 feet. Of these three bridges only London bridge has survived to nowaday. Perhaps the most extraordinary thing about it is its great cost of £1,458,311, an immense sum in those days for a multi-span bridge little more than 1,000 feet long. In addition to his fame as a bridge builder Rennie had made a reputation for himself in successfully draining and reclaiming marshes in the eastern counties and in building a number of harbours and docks including the London dock and East India dock on the Thames.

Before closing this article we must refer to an outstanding example of intuitive skill in construction shown by two village carpenters, Ulric and Jean Grubenmann of Teufen, Switzerland. Their most famous work was the covered timber bridge over the Rhine at Schaffhausen built in 1755—1758. This bridge had two spans of 193 feet and 171 feet respectively, which formed an elbow pointing upstream. Ulric had proposed to build a single feet span across the river, but the town magistrates insisted that he should make use of a remaining pier in midstream. The bridge was completely successful and cost only £8,000. Unfortunately it was burnt by French troops when they evacuated Schaffhausen after being defeated by the Austrians, in 1799. Telford made a report on the bridge in which he agreed that if it had been built straight across the river it might well have stood as a single span.

Before the work was finished, Jean Grubenmann built a similar timber bridge of 240 feet span at Reichenau. By this time the fame of the brothers Grubenmann had travelled to North America, where a great variety of timber truss bridges were evolved. Foremost among the designers of these were Theodore Burr, Louis Wernwag, who built the Colossus bridge of 340-feet span over the Schuylkill river, and William Howe. The majority of the designs were indeterminate, however, in that they combined arch and truss systems. The most satisfactory was the Howe truss, a purely lattice bridge with vertical iron ties, which was largely used in the early railroads. Although it has many valuable properties, however, timber is not a suitable material for permanent bridgework and would not be used in work of importance today.

课文选自：苏小卒. 土木工程专业英语（下册）[M]. 上海：同济大学出版社，2018.

阅读选自：苏小卒. 土木工程专业英语（下册）[M]. 上海：同济大学出版社，2018.

Lesson 6 :
Pavement

A highway pavement is a structure consisting of superimposed layers of selected and processed materials placed on a subgrade, whose primary function is to support the applied traffic loads and distribute them to the basement soil. The ultimate aim is to ensure that the transmitted stresses are sufficiently reduced that they will not exceed the supporting capacity of the subgrade. Two types of pavement are arbitrarily recognized as serving this purpose — flexible pavements and rigid pavements.

A flexible pavement is a pavement structure that maintains intimate contact with, and distributes loads to, the subgrade. It depends upon aggregate interlock, particle friction, and cohesion for its stability. The distinguishing feature of flexible pavement lies in its structural mechanics — the pressure is usually assumed to be transmitted to the subgrade through the lateral distribution of the applied load with depth, rather than by beam and slab action as with a concrete slab. Thus a flexible pavement can be most easily defined by contrasting it with a rigid Portland cement concrete pavement.

When the subgrade deflects beneath a rigid pavement, the concrete slab is able to bridge over localized failures and areas of inadequate support because of its rigidity and high modulus of elasticity. The major factor influencing the design of a rigid pavement is the structural strength of the concrete, and its thickness is relatively less affected by the quality of the subgrade as long as it meets certain minimum criteria.

In direct contrast to this, the strength of the subgrade is a major factor controlling the design of a flexible pavement. When the subgrade deflects, the overlying flexible pavement is assumed to deform to a similar shape and extent. In fact, this does not necessarily happen, e.g, pavements with bituminous-bound or chemically stabilized roadbases have beam strengths which increase with thickness and help support the imposed loads. Nonetheless, these pavements are still generally classed as flexible pavements, and the assumed basic design criterion is that a depth of pavement is required that will distribute the applied surface load through the various pavement layers to the subgrade so that the subgrade is not over-stressed.

In its simplest form, a flexible pavement is generally considered to be any pavement other than a concrete one. It is this definition that is accepted by the great majority of practicing engineers. It should be clearly understood, however, that the term is simply one of convenience and does not truly reflect the characteristics of the many different and composite types of construction masquerading as "flexible" pavements.

Whether it is a pavement for an expensive motorway or a simple country road, the basic structural cross-section of a flexible road is essentially that illustrated in Fig. 6-1, it is composed of several distinct layers that make up the pavement superimposed on the subgrade in the manner indicated. The intersection of the subgrade and the pavement is known as the formation.

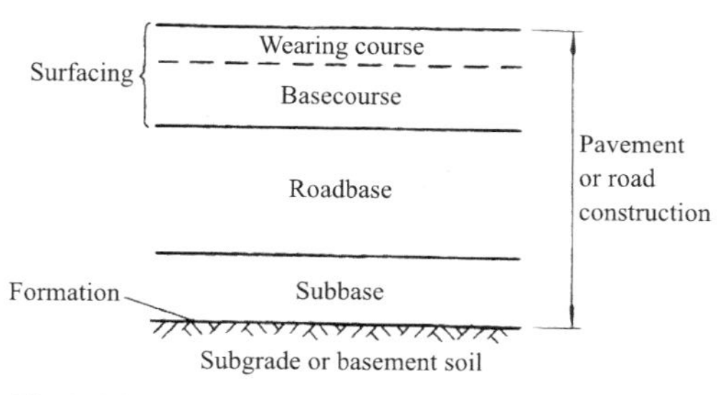

Fig. 6-1 Basic structural cross-section of a flexible road

The subgrade is normally considered to be the in situ soil over which the highway is being constructed. It should be quite clear, however, that the term subgrade is also applied to all native soil materials exposed by excavation and to excavated soil that may be artificially deposited to form a compacted embankment. In the latter case, the added material is not considered to be part of the road structure itself but part of the foundation of the road.

The uppermost layer of a flexible pavement is called the surface course. The highway materials used in a surface course can vary from loose mixtures of soil and gravel to the very-highest-quality bituminous mixtures. The choice of materials used in any particular situation depends in most countries upon the quality of service required of the highway.

If a surface course is composed of bituminous materials — as is the normal practice for flexible pavements in Britain — it may consist of a single homogeneous layer or, in the higher-quality roads, of two distinct sub-layers known as a wearing course and a basecourse. The wearing course provides the actual surfacing on which the vehicles run, whilst the basecourse acts as a regulating layer to provide the wearing course with a better riding quality. The basecourse is normally composed of a more pervious material than the wearing course.

The primary function of the surface course, and especially of its wearing course component, is to provide a safe and comfortable riding surface for traffic. It must also withstand the most concentrated stresses due to traffic, and protect the pavement layers beneath from the effects of the natural elements.

Bituminous surfacings are generally expected: (a) to contribute to the structural strength of the pavement, (b) to provide a high resistance to plastic deformation and resistance to cracking under traffic, and (c) to maintain such desirable surface characteristics as good skid-resistance, good drainage, and low tyre noise.

Roadbase must not be confused with the basecourse which is an integral part of the surface course. One is a sub-layer within the bituminous surfacing, while the other is normally the thickest element of the flexible pavement on which the surfacing rests.

From a structural aspect, the roadbase is the most important layer of a flexible pavement. It is expected to bear the burden of distributing the applied surface loads so that the bearing capacity of the subgrade is not exceeded. Since it provides the pavement with added stiffness and resistance to fatigue, as well as contributing to the overall thickness, the material used in a roadbase must always be of a reasonably high quality. Roadbase materials rang from unbound soils and/or aggregates, to chemically stabilized soils, to cement/bitumen-bound materials.

In its simplest sense, subbase can be considered merely as an extension of the roadbase. In fact, it may or may not be present in the pavement as a separate layer. Whether or not it is utilized in a pavement depends upon the purpose for which it is to be used. Its function can be examined from a number of aspects, as follows.

(1) As a structural member of the pavement the subbase helps to distribute the applied loads to the subgrade. The subbase material must always be significantly stronger than the subgrade material and capable of resisting within itself the stresses transmitted to it via the roadbase.

(2) A coarse-grained material may be used in the subbase to act as a drainage layer, i. e.to pass to the highway drainage system any moisture which falls during construction or which enters the pavement after construction. The quality of the material used must be such that the free-draining criterion of the subbase is always met. In certain instances, this may require a dual-layer subbase, i. e. an open-graded layer with a protective filter.

(3) On fine-grained subgrade soils a granular subbase may be provided: (a) to carry constructional traffic and act as a working platform on which subsequent layers can be constructed, (b) to act as a cutoff blanket and prevent moisture from migrating upward from the subgrade, or (c) to act as a cutoff blanket to prevent the infiltration of subgrade material into the pavement.

The type of material used in any of these designs depends upon the purpose for which it being used and the grading of the subgrade soil.

1. Phrases and Expressions

flexible (rigid) pavement　柔性（刚性）路面
formation　|fɔːrˈmeɪʃn|　*n.* 路基面
regulating layer　整平层
roadbase　|ˌroʊdˈbeɪs|　*n.* 道路基层
subbase　|ˈsʌbˌbeɪs|　*n.* 底基层，副基层

subgrade |ˈsʌbˌgreɪd| *n.* 路基，地基
surface course 面层

2. Please translate the following sentences into Chinese

(1) A highway pavement is a structure consisting of superimposed layers of selected and processed materials placed on a subgrade, whose primary function is to support the applied traffic loads and distribute them to the basement soil.

(2) A flexible pavement is a pavement structure that maintains intimate contact with, and distribute loads to, the subgrade. It depends upon aggregate interlock, particle friction, and cohesion for its stability.

(3) The major factor influencing the design of a rigid pavement is the structural strength of the concrete, and its thickness is relatively less affected by the quality of the subgrade as long as it meets certain minimum criteria.

(4) The basic structural cross-section of a flexible road is composed of several distinct layers that make up the pavement superimposed on the subgrade in a certain manner.

(5) The primary function of the surface coarse, and especially of its wearing course component, is to provide a safe and comfortable riding surface for traffic.

3. Please translate the following sentences into English

（1）铺设路面的最终目的是使其所传递的应力降到最低、不超过路基的承载能力。

（2）按照表面荷载的传递方式，路面可分为柔性路面和刚性路面。

（3）柔性路面的明显特征是它的传力路径：荷载大致按均布荷载沿路面厚度传到路基，而不是像框架结构中的混凝土板一样将荷载以集中力的形式传递到梁上。

（4）由于刚性路面具有较大刚度和高弹性模量，所以它能将荷载传布到较大面积的路基土壤上。

（5）从结构的角度讲，基层是柔性路面最重要的一层，因为路面荷载是依靠它传布到路基上的。

4. Extension

acoustic barrier　隔声墙
alignment design　（城市道路）平面设计，线形设计
alignment element　线形要素
allowable rebound deflection　容许（回弹）弯沉
average gradient　平均纵坡
bicycle path　自行车道
bituminous concrete pavement　沥青混凝土路面
climbing lane　爬坡车道
convex vertical curve　凸形竖曲线
cross slope　横坡
cross walk　人行横道
curb　|kɜːrb|　*n.* 路缘石
cut-fill transition program　土方调配图
cutting　|ˈkʌtɪŋ|　*n.* 路堑
deflection test　弯沉试验
degree of compaction　压实度
design of elevation　（城市道路）竖向设计
design of vertical alignment　纵断面设计
earth bank　路堤
emergency parking strip　紧急停车带
expressway　|ɪkˈspresweɪ|　*n.*（城市）快速路
highway　|ˈhaɪweɪ|　*n.* 公路
loop ramp　环形匝道
overtaking lane　超车车道
route selection　选线
side trip　路缘带
side ditch　边沟
side slope　边坡

traffic lane 车道
shoulder of road 路肩

5. Reading material

Bituminous Materials and Pavements

According to nomenclature commonly in use in the United States, the term "bituminous material" is used to denote substances in which bitumen is present or from which it can be derived. Bitumen is hydrocarbon material of either natural or pyrogenous origin, gaseous, liquid, semisolid, or solid, which is completely soluble in carbon disulfide (ASTM) .

With respect to use in highway construction, the term bituminous material is used to include both natural and manufactured materials regardless of origin, but is restricted to those hydrocarbon materials which are cementitious in character or from which a residuum of this character will develop.

Terms Relating to Asphalt

Asphalts: Black to dark brown semisolid to solid cementitious materials consisting principally of bitumen that gradually liquefy when heated and which occur in nature as such or are obtained as a residuum in the refining of petroleum.

Asphalt Cement, or Paving Asphalt: An asphalt specially prepared as to quality and consistency for direct use in paving. It has a normal penetration between 40 and 300 and must be used in hot condition.

Native, or Natural Asphalt: One occurring as such in nature. It may be of the lake, rock, or vein variety, may be essentially pure bitumen or contain a large amount of mineral matter, and the asphalt may vary from hard to soft.

Petroleum Asphalt: Asphalt produced by the refining of petroleum.

Cutback Asphalt: Asphalt dissolved in naphtha (gasoline) or kerosene to render it temporarily fluid for use. If the solvent is naphtha and dissolved asphalt relatively hard, the cutback is known as a rapid-curing one. If the solvent is kerosene and the dissolved asphalt relatively soft, the cutback is designated as a medium-curing one.

Emulsified Asphalt: A mixture in which an asphalt cement in a finely dispersed state is suspended in chemically treated water. An inverted asphalt emulsion is one in which asphalt is the continuous phase with water dispersed in it. Inverted asphalt emulsions used in road construction usually are made from liquid asphaltic materials.

Terms Relating to Bituminous Pavements

Bituminous Pavements: Highway or similar structures, placed to carry traffic, which consist of a mineral-aggregate surface cemented together with bituminous material. The bituminous portion may be only a thin surface, or it may be 8 in. (20 cm) or more in thickness, consisting of mixtures of aggregate and bituminous material more or less carefully proportioned and compacted upon a foundation soil, gravel or stone base, or old bituminous, cement concrete, brick, or block pavement.

Prime Coat: Liquid bituminous material applied to penetrate the surface of a base on which a bituminous pavement is to be placed. The prime coat is applied to bond together loose particles on the surface, to protect the surface from weather, traffic and construction equipment, and to prepare the surface to receive the bituminous pavement.

Tack Coat: A bituminous material applied to provide bond between a new or old surface and the mixture with which it is being covered. It may be used in addition to a prime coat and may be applied as a hot asphalt cement or as a liquid bituminous material.

Seal Coat: An application of bituminous material, usually with aggregate cover and seldom over 1/2 in. (1.3 cm) in thickness, which is applied for one or more of the following reasons: (1) to prevent the entrance of air or moisture, (2) to rejuvenate an old, dry, or weathered surface, (3) to provide a nonskid surface texture, (4) to change surface color for visibility or for demarcation purposes, and (5) to supply additional asphalt on the surface for more effective sealing by traffic.

Bituminous Base Course: Consists of mineral aggregate such as stone, gravel, or sand bonded together by a bituminous material and used as a foundation upon which to place a binder or surface course.

Bituminous Binder Course: A bituminous-aggregate mixture used as an intermediate course between the base and surface courses or as the first bituminous layer in two-course bituminous resurfacing. It is sometimes called a leveling course, if that is its primary function.

Bituminous Surface Course: The uppermost course of a bituminous pavement.

Bituminous Surface Treatment: The application of mineral aggregate and bituminous material to a road or pavement surface, either with or without a mixing operation, which results in a surface course not substantially greater than 1 in. (2.5 cm) in thickness. Double-or triple-surface treatments are constructed with multiple applications of bituminous material and aggregate.

Penetration Macadam: A type of bituminous-pavement construction in which relatively coarse and uniform-size stone aggregate is placed in layers and stabilized by rolling sufficiently to cause an interlocking of the aggregate, after which the course is penetrated by bituminous material. The bituminous material may be asphalt cement or a heavy tar applied hot or rapid-curing cutback or rapid-setting emulsion.

Mixed-in-Place, or Road Mix: A method of mixing mineral aggregate and bituminous material on the road surface by means of blade graders, drags, farm implements, or special road-mixing equipment.

Travel-Plant Mix: Refers to a method of mixing aggregate and bituminous material in which a traveling mechanical mixer is used. The aggregate may be in place in a windrow or may be dumped into the hopper of the traveling mixer by trucks.

Plant Mix: A term which refers to the mixing of mineral aggregate and bituminous material in a mechanical mixer into which the aggregates and bituminous material are carefully proportioned. The mixture may be hot or cold, and cold mixes may or may not be produced with dried aggregate. The resulting mix is transported to the road and laid either hot or cold.

Cold-Laid Mixtures: Plant mixes made with bituminous materials of the liquid type or with solvents or flux oils of a type that will permit the mixture to be spread and compacted at atmospheric temperature. Some cold-laid mixtures are produced hot.

Asphalt Concrete: A plant mix of closely graded mineral aggregate and asphalt, designed and controlled to produce a mixture of high quality from the standpoint of both stability and durability. Such mixtures are usually produced hot with asphalt cement, but other types may be used as long as the resulting mixture is as described. It may be produced as base, binder, or surface courses.

Sheet Asphalt: A hot plant mixture of sand, filler and asphalt cement carefully proportioned and controlled to produce a sand-type mixture of high quality. It is usually used for surfacing courses.

Sand Asphalt: A mixture of sand and asphalt that may be made with asphalt cement, emulsified asphalt, or other liquid type, mixed in a plant or by road-mix methods.

课文选自：李嘉. 专业英语（公路、桥梁工程专用）[M]. 北京：人民交通出版社，2004.
阅读选自：李嘉. 专业英语（公路、桥梁工程专用）[M]. 北京：人民交通出版社，2004.

Lesson 7 :

Geotechnical Engineering

Geotechnical engineering, also known as geotechnics, is the branch of civil engineering concerned with the engineering behavior of earth materials. It uses the principles of soil mechanics and rock mechanics for the solution of its respective engineering problems. It also relies on knowledge ofgeology, hydrology, geophysics and other related sciences. Geotechnical (rock) engineering is a subdiscipline of geological engineering. In addition to civil engineering, geotechnical engineering also has applications in military, mining, petroleum, coastal engineering and offshore construction. The fields of geotechnical engineering and engineering geology have knowledge areas that overlap, however, while geotechnical engineering is a specialty of civil engineering, engineering geology is a specialty of geology: They share the same principles of soil mechanics and rock mechanics, but differ in the application.

Humans have historically used soil as a material for flood control, irrigation purposes, burial sites, building foundations, and as construction material for buildings. First activities were linked to irrigation and flood control, as demonstrated by traces of dykes, dams and canals dating back to at least 2000 BCE that were found in ancient Egypt, ancient Mesopotamia and the Fertile Crescent, as well as around the early settlements of Mohenjo Daro and Harappa in the Indus valley. As the cities expanded, structures were supported by formalized foundations. Ancient Greeks notably constructed pad footings and strip-and-raft foundations. Until the 18th century, however, no theoretical basis for soil design had been developed and the discipline was more of an art than a science, relying on past experience.

Several foundation-related engineering problems, such as the Leaning Tower of Pisa, prompted scientists to begin taking a more scientific-based approach to examining the subsurface. The earliest advances occurred in the development of earth pressure theories for the construction of retaining walls. Henri Gautier, a French Royal Engineer, recognized the "natural slope" of different soils in 1717, an idea later known as the soil' s angle of repose. A rudimentary soil classification system was also developed based on a material' s unit weight, which is no longer considered a good indication of soil type.

The application of the principles of mechanics to soils was documented as early as 1773 when Charles Coulomb (a physicist, engineer and army Captain) developed improved methods to determine the earth pressures against military ramparts. Coulomb observed that, at failure, a distinct slip plane would form behind a sliding retaining wall and he suggested that the maximum shear stress on the slip plane, for design purposes, was the sum of the soil cohesion, c, and friction σ tan(φ), where σ, is the normal stress on the slip plane and φ is the friction angle of the soil. By combining Coulomb' s theory with Christian Otto Mohr' s 2D stress state, the theory became known as Mohr-Coulomb theory. Although it is now recognized that precise determination of cohesion is impossible because c is not a fundamental soil property, the Mohr-Coulomb theory is still used in practice today.

In the 19th century Henry Darcy developed what is now known as Darcy' s Law describing the flow of fluids in porous media. Joseph Boussinesq (a mathematician and physicist) developed theories of stress distribution in elastic solids that proved useful for estimating stresses at depth in the ground. William Rankine, an engineer and physicist, developed an alternative to Coulomb' s earth pressure theory. Albert Atterberg developed the clay consistency indices that are still used today for soil classification. Osborne Reynolds recognized in 1885 that shearing causes volumetric dilation of dense and contraction of loose granular materials.

Modern geotechnical engineering is said to have begun in 1925 with the publication of Erdbaumechanik by Karl Terzaghi (a mechanical engineer and geologist) . Considered by many to be the father of modern soil mechanics and geotechnical engineering, Terzaghi developed the principle of effective stress, and demonstrated that the shear strength of soil is controlled by effective stress. Terzaghi also developed the framework for theories of bearing capacity of foundations, and the theory for prediction of the rate of settlement of clay layers due to consolidation. Afterwards, Maurice Biot fully developed the three-dimensional soil consolidation theory, extending the one-dimensional model previously developed by Terzaghi to more general hypotheses and introducing the set of basic equations of Poro-elasticity. Alec Skempton in his work in 1960, has carried out an extensive review of available formulations and experimental data in literature about effective stress valid in soil, concrete and rock, in order to reject some of these expressions, as well as clarify what expression was appropriate according to several work hypotheses, such as stress-strain or strength behavior, saturated or nonsaturated media, rock/concrete or soil behavior, etc. In his 1948 book, Donald Taylor recognized that interlocking and dilation of densely packed particles contributed to the peak strength of a soil. The interrelationships between volume change behavior (dilation, contraction and consolidation) and shearing behavior were all connected via the theory of plasticity using critical state soil mechanics by Roscoe, Schofield, and Wroth with the publication of "On the Yielding of Soils" in 1958. Critical state soil mechanics is the basis for many contemporary advanced constitutive models describing the behavior of soil.

Geotechnical centrifuge modeling is a method of testing physical scale models of geotechnical problems. The use of a centrifuge enhances the similarity of the scale model tests involving soil because the strength and stiffness of soil is very sensitive to the confining pressure. The scale model is typically constructed in the laboratory and then loaded onto the end of the centrifuge, which is typically between 0.2 and 10 meters in radius. The purpose of spinning the models on the centrifuge is to increase the g-forces on the model so that stresses in the model are equal to stresses in the prototype. For example, the stress beneath a 0.1-metre-deep layer of model soil spun at a centrifugal acceleration of 50 g produces stresses equivalent to those beneath a 5-metre-deep prototype layer of soil in earth's gravity. The idea to use centrifugal acceleration to simulate increased gravitational acceleration was first proposed by Phillips in 1869. During the early stage of 1930s, Pokrovsky and Fedorov in the Soviet Union and Bucky in the United States were the first to implement the idea. In 1980, Andrew N. Schofield played a key role in modern development of centrifuge modeling. Overall, the centrifuge applies an increased "gravitational" acceleration to physical models in order to produce identical self-weight stresses in the model and prototype. In addition, the centrifugal acceleration allows a researcher to obtain large (prototype-scale) stresses in small physical models.

1. Phrases and Expressions

soil mechanics *n.* 土壤力学
geological engineering *n.* 地质工程
strip-and-raft foundations *n.* 条筏基础
earth pressures *n.* 土压力
soil cohesion *n.* 土壤黏聚力
friction coefficient *n.* 摩擦系数
mohr-coulomb strength theory *n.* 摩尔 - 库伦强度理论
Darcy's Law *n.* 达西定律
elastic |ɪˈlæstɪk| *adj.* 弹性的
bearing capacity *n.* 承载力
rate of settlement *n.* 沉降速率
poro-elasticity |poʊroʊelæsˈtɪsɪtɪ| *n.* 孔隙弹性理论
stress–strain *n.* 应力应变
dilation |daɪˈleɪʃn| *n.* 膨胀
contraction |kənˈtrækʃn| *n.* 收缩
consolidation |kənˌsɑːlɪˈdeɪʃn| *n.* 固结

2. Please translate the following sentences into Chinese

(1) Geotechnical engineering, also known as geotechnics, is the branch of civil engineering concerned with the engineering behavior of earth materials.

(2) Humans have historically used soil as a material for flood control, irrigation purposes, burial sites, building foundations, and as construction material for buildings.

(3) Considered by many to be the father of modern soil mechanics and geotechnical engineering, Terzaghi developed the principle of effective stress, and demonstrated that the shear strength of soil is controlled by effective stress.

(4) Afterwards, Maurice Biot fully developed the three-dimensional soil consolidation theory, extending the one-dimensional model previously developed by

Terzaghi to more general hypotheses and introducing the set of basic equations of Poro-elasticity.

(5) The interrelationships between volume change behavior (dilation, contraction and consolidation) and shearing behavior were all connected via the theory of plasticity using critical state soil mechanics.

3. Please translate the following sentences into English

（1）岩土工程，是土木工程的一个分支，该工程领域利用土力学和岩石力学的原理来解决工程问题。

（2）滑动挡土墙在破坏时会形成一个明显的滑移面，其滑面上的最大剪应力是土壤黏聚力和摩擦力的总和。

（3）现代岩土工程中发展了有效应力原理，并证明了土壤的抗剪强度受有效应力的控制。

（4）基于地基承载力理论框架，工程师可以预测由于固结引起的黏土层沉降率。

（5）临界状态土力学的塑性理论解释了体积变化过程与土的剪切行为之间的关系。

4. Extension

irrigation |ˌɪrɪˈgeɪʃn| *n.* 灌溉
dykes |daɪks| *n.* 堤
dams |dæmz| *n.* 水坝
canals |kəˈnælz| *n.* 运河
soil classification *n.* 土壤分类
granular materials *n.* 粒状材料
non-saturated porous media *n.* 非饱和孔隙介质

5. Reading material

Soil Consolidation

Soil consolidation refers to the mechanical process by which soil changes volume gradually in response to a change in pressure. This happens because soil is a two-phase material, comprising soil grains and pore fluid, usually groundwater. When soil saturated with water is subjected to an increase in pressure, the high volumetric stiffness of water compared to the soil matrix means that the water initially absorbs all the change in pressure without changing volume, creating excess pore water pressure. As water diffuses away from regions of high pressure due to seepage, the soil matrix gradually takes up the pressure change and shrinks in volume. The theoretical framework of consolidation is therefore closely related to the diffusion equation, the concept of effective stress, and hydraulic conductivity.

In the narrow sense, "consolidation" refers strictly to this delayed volumetric response to pressure change due to gradual movement of water. Some publications also use "consolidation" in the broad sense, to refer to any process by which soil changes volume due to a change in applied pressure. This broader definition encompasses the overall concept of soil compaction, subsidence and heave. Some types of soil, mainly those rich in organic matter, show significant creep, whereby the soil changes volume slowly at constant effective stress over a longer time-scale than consolidation due to the diffusion of water. To distinguish between the two mechanisms, "primary consolidation" refers to consolidation due to dissipation of excess water pressure, while "secondary consolidation" refers to the creep process.

The effects of consolidation are most conspicuous where a building sits over a layer of soil with low stiffness and low permeability, such as marine clay, leading to large settlement over many years. Types of construction project where consolidation often poses technical risk include land reclamation, the construction of embankments, and tunnel and basement excavation in clay.

Geotechnical engineers use oedometers to quantify the effects of consolidation. In an oedometer test, a series of known pressures are applied to a thin disc of soil sample, and the change of sample thickness with time is recorded. This allows the consolidation characteristics of the soil to be quantified in terms of the coefficient of consolidation (C_v) and hydraulic conductivity (K).

According to the "father of soil mechanics", Karl von Terzaghi, consolidation is “any process which involves a decrease in water content of saturated soil without replacement of water by air”. More generally, consolidation refers to the process by which soils change volume in response to a change in pressure, encompassing both compaction and swelling.

When stress is applied to a soil, it causes the soil particles to pack together more tightly. When this occurs in a soil that is saturated with water, water will be squeezed out of the soil. The magnitude of consolidation can be predicted by many different methods. In the classical method developed by Terzaghi, soils are tested with an oedometer test to determine their compressibility. When stress is removed from a consolidated soil, the soil will rebound, regaining some of the volume it had lost in the consolidation process. If the stress is reapplied, the soil will consolidate again along a recompression curve, defined by the recompression index.

The soil which had its load removed is considered to be “overconsolidated”. This is the case for soils that have previously had glaciers on them. The highest stress that it has been subjected to is termed the “preconsolidation stress”. The “over-consolidation ratio” (OCR) is defined as the highest stress experienced divided by the current stress. A soil that is currently experiencing its highest stress is said to be “normally consolidated” and has an OCR of one. A soil could be considered “underconsolidated” or “unconsolidated” immediately after a new load is applied but before the excess pore water pressure has dissipated. Occasionally, soil strata form by natural deposition in rivers and seas may exist in an exceptionally low density that is impossible to achieve in an oedometer, this process is known as “intrinsic consolidation”.

课文选自： Das B M. Principles of geotechnical engineering[M]. Cengage learning, 2006.

阅读选自： Lambe T W, Whitman R V. Soil mechanics[M]. John Wiley & Sons, 1991.

Lesson 8 :

Geotechnical Investigation

Geotechnical investigations are performed by geotechnical engineers or engineering geologists to obtain information on the physical properties of soil earthworks and foundations for proposed structures and for the repair of distress to earthworks and structures caused by subsurface conditions. This type of investigation is called a site investigation. The tasks of a geotechnical engineer comprise the investigation of subsurface conditions and materials; the determination of the relevant physical, mechanical and chemical properties of these materials; the design of earthworks and retaining structures (including dams, embankments, sanitary landfills, deposits of hazardous waste), tunnels and structure foundations; the monitoring of site conditions, earthwork and foundation construction; the evaluation of the stability of natural slopes and man-made soil deposits; the assessment of the risks posed by site conditions; and the prediction, prevention and mitigation of damage caused by natural hazards (such as avalanches, mudflows, landslides, rockslides, sinkholes and volcanic eruptions) .

Geotechnical engineers and engineering geologists perform geotechnical investigations to obtain information on the physical properties of soil and rock underlying (and sometimes adjacent to) a site to design earthworks and foundations for proposed structures, and for the repair of distress to earthworks and structures caused by subsurface conditions. A geotechnical investigation will include surface exploration and subsurface exploration of a site. Sometimes, geophysical methods are used to obtain data about sites. Subsurface exploration usually involves in-situ testing (two common examples of in-situ tests are the standard penetration test and cone penetration test) . In addition, site investigation will often include subsurface sampling and laboratory testing of the soil samples retrieved. The digging of test pits and trenching (particularly for locating faults and slide planes) may also be used to learn about soil conditions at depth. Large diameter borings are rarely used due to safety concerns and expense but are sometimes used to allow a geologist or engineer to be lowered into the borehole for direct visual and manual examination of the soil and rock stratigraphy.

In terms of soil sampling, borings come in two main varieties, including the large-diameter and the small-diameter. Large-diameter borings are rarely used due to safety concerns and expense but are sometimes used to allow a geologist or an engineer to visually and manually examine the soil and rock stratigraphy in-situ. Small-diameter borings are frequently used to allow a geologist or engineer to examine soil or rock cuttings or to retrieve samples at depth using soil samplers, and to perform in-place soil tests.

Soil samples are often categorized as being either disturbed or undisturbed, however, "undisturbed" samples are not truly undisturbed. A disturbed sample is one in which the structure of the soil has been changed sufficiently that tests of structural properties of the soil will not be representative of in-situ conditions, and only properties of the soil grains (e.g., grain size distribution, Atterberg limits, compaction characteristic of soil, to determine the general lithology of soil deposits and possibly the water content) can be accurately determined. An undisturbed sample is one where the condition of the soil in the sample is close enough to the conditions of the soil in-situ to allow tests of structural properties of the soil to be used to approximate the properties of the soil in-situ. Specimen obtained by the undisturbed method determines the soil stratification, permeability, density, consolidation, and other engineering characteristics.

On the other hand, offshore soil collection introduces many difficult variables. In shallow water, work can be done off a barge. In deeper water, a ship will be required. Deepwater soil samplers are normally variants of Kullenberg-type samplers, a modification on a basic gravity corer using a piston. Seabed samples are also available, which push the collection tube slowly into the soil.

A variety of soil samplers exists to meet the needs of different engineering projects. The standard penetration test (SPT), which uses a thick-walled split spoon sampler, is the most common way to collect disturbed samples. Piston samplers, employing a thin-walled tube, are most commonly used for the collection of less disturbed samples. Coring frozen ground provides high-quality undisturbed samples from any ground conditions, such as fill, sand, moraine and rock fracture zones.

A wide variety of laboratory tests can be performed on soils to measure a wide variety of soil properties. Some soil properties are intrinsic to the composition of the soil matrix and are not affected by sample disturbance, while other properties depend on the structure of the soil as well as its composition, and can only be effectively tested on relatively undisturbed samples. Some soil tests measure the direct properties of the soil, while others measure "index properties", which provide useful information about the soil without directly measuring the property desired. Among them, Atterberg limits tests, water content measurements, and grain size analysis may be performed on disturbed samples obtained from thick-walled soil samplers. Properties such as shear strength, stiffness, hydraulic conductivity, and coefficient of consolidation may be significantly altered by sample disturbance. To measure these properties in the laboratory, high-quality sampling is required. Common tests to measure strength and stiffness include the triaxial shear and unconfined compression test.

Surface exploration can include geologic mapping, geophysical methods, and photogrammetry, or it can be as simple as an engineer walking around to observe the physical conditions at the site. first of all, a geologic map or geological map is a special-purpose map made to show various geological features. For example, rock units or geologic strata are indicated by color or symbols. Bedding planes and structural features such as faults and folds are shown with strike and dip or trend and plunge symbols, which give three-dimensional orientation features. Secondly, exploration geophysics is an applied branch of geophysics and economic geology, which uses physical methods, such as seismic, gravitational, magnetic, electrical and electromagnetic at the surface of the Earth to measure the physical properties of the subsurface, along with the anomalies in those properties. It is most often used to detect or infer the presence and position of economically useful geological deposits, such as ore minerals, fossil fuels and other hydrocarbons, geothermal reservoirs and groundwater reservoirs. Last but not least, photogrammetry is the science and technology of obtaining reliable information about physical objects and the environment through the process of recording, measuring and interpreting photographic images and patterns of electromagnetic radiant imagery and other phenomena.

1. Phrases and Expressions

subsurface |ˈsʌbˌsɜːrfɪs| *adj.* 地下的
embankment |ɪmˈbæŋkmənt| *n.* 路堤
sanitary landfills *n.* 卫生填埋
hazardous waste *n.* 有害垃圾
man-made |ˌmæn ˈmeɪd| *adj.* 人造的
mudflow |ˈmʌdˌfloʊ| *n.* 泥石流
landslides |ˈlændslaɪds| *n.* 滑坡
exploration |ˌekspləˈreɪʃn| *n.* 勘探
standard penetration test *n.* 标准贯入试验
cone penetration test *n.* 静力触探试验
soil samples *n.* 土壤样本
earthwork |ˈɜːrθwɜːrk| *n.* 土方工程
mitigation |ˌmɪtɪˈgeɪʃn| *n.* 减轻
in-situ |ˌɪn ˈsaɪtuː| *adj.* 现场的

2. Please translate the following sentences into Chinese

(1) Geotechnical engineers and engineering geologists perform geotechnical investigations to obtain information on the physical properties of soil and rock underlying a site to design earthworks and foundations for proposed structures.

(2) A geotechnical investigation will include surface exploration and subsurface exploration of a site.

(3) In addition site investigation will often include subsurface sampling and laboratory testing of the soil samples retrieved.

(4) Properties such as shear strength, stiffness, hydraulic conductivity, and coefficient of consolidation may be significantly altered by sample disturbance.

(5) Surface exploration can include geologic mapping, geophysical methods, and photogrammetry, or it can be as simple as an engineer walking around to observe the physical conditions at the site.

3. Please translate the following sentences into English

（1）岩土工程师的任务其中包括评估现场条件带来的风险，以及预测、预防和减轻自然灾害（如雪崩、泥石流、山体滑坡、岩崩、塌陷和火山爆发）造成的损害。

（2）通过进行岩土工程调查，可以获得场地下方土壤和岩石的物理特性，用于设计土方工程和结构的基础。

（3）此外，现场调查常常包括地下取样和对所获得的土壤样品进行实验室测试。

（4）现有多种土壤取样器，满足不同工程项目的需要，包括标准贯入试验取样器、活塞取样器、冻土取样器等。

（5）诸如抗剪强度、刚度、水力传导率和固结系数等特性可能会因试样扰动而显著改变。

4. Extension

slide plane *n.* 滑动面
borehole |ˈbɔːrhoʊl| *n.* 钻孔
fracture zone *n.* 断裂带
stratigraphy |strəˈtɪgrəfi| *n.* 地层学
atterberg limits test *n.* 界限含水量试验
triaxial shear test *n.* 三轴剪切试验
unconfined compression test *n.* 无侧限压缩试验
avalanche |ˈævəlæntʃ| *n.* 雪崩
rockslide |ˈrɑːkˌslaɪd| *n.* 岩滑
sinkhole |ˈsɪŋkhoʊl| *n.* 沉陷坑

5. Reading material

Soil Compaction

In geotechnical engineering, soil compaction is the process in which stress applied to a soil causes densification as air is displaced from the pores between the soil grains. When stress is applied that causes densification due to water (or other liquid) being displaced from between the soil grains, then consolidation, not compaction, has occurred. Normally, compaction is the result of heavy machinery compressing the soil, but it can also occur due to the passage of, for example, animal feet.

In soil science and agronomy, soil compaction is usually a combination of both engineering compaction and consolidation, so it may occur due to a lack of water in the soil, the applied stress being internal suction due to water evaporation, and due to passage of animal feet. Affected soils become less able to absorb rainfall, thus increasing runoff and erosion. Plants have difficulty compacting soil because the mineral grains are pressed together, leaving little space for air and water, essential for root growth. Burrowing animals also find it a hostile environment because the denser soil is more difficult to penetrate. The ability of soil to recover from this type of compaction depends on climate, mineralogy and fauna. Soils with high shrink-swell capacity, such as vertisols, recover quickly from compaction where moisture conditions are variable (dry spells shrink the soil, causing it to crack) . But clays such as kaolinite, which do not crack as they dry, cannot recover from compaction on their own unless they host ground-dwelling animals such as earthworms. Before soils can be compacted in the field, some laboratory tests are required to determine their engineering properties. Among various properties, the maximum dry density and the optimum moisture content are vital and specify the required density to be compacted in the field.

Soil compaction is a vital part of the construction process. It is used for support of structural entities such as building foundations, roadways, walkways and earth retaining structures, to name a few. Certain properties may deem it more or less desirable to perform adequately for a particular circumstance for a given soil type. In general, the preselected soil should have adequate strength, be relatively incompressible so that future settlement is not significant, be stable against volume change as water content or other factors vary, be durable and safe against deterioration, and possess proper permeability.

When an area is to be filled or backfilled, the soil is placed in layers called lifts. The ability of the first fill layers to be properly compacted will depend on the condition of the natural material being covered. If unsuitable material is left in place and backfilled, it may compress over a long period under the weight of the earth fill, causing settlement cracks in the fill or in any structure supported by the fill. In order to determine if the natural soil will support the first fill layers, an area can be proof rolled. Proofrolling consists of utilizing heavy construction equipment to roll across the fill site and watching for deflections to be revealed. These areas will be indicated by the development of rutting, pumping, or ground weaving. To ensure adequate soil compaction is achieved, project specifications will indicate the required soil density or degree of compaction that must be achieved. A geotechnical engineer generally recommends these specifications in a geotechnical engineering report.

The soil type-grain-size distributions, the shape of the soil grains, the specific gravity of soil solids, and the amount and type of clay minerals present — greatly influences the maximum dry unit weight and optimum moisture content. It also greatly influences how the materials should be compacted in given situations. Compaction is accomplished by the use of heavy equipment. In sands and gravels, the equipment usually vibrates to cause re-orientation of the soil particles into a denser configuration. A sheepsfoot roller is frequently used in silts and clays to create small zones of intense shearing, which drives air out of the soil.

Determination of adequate compaction is done by determining the in-situ density of the soil and comparing it to the maximum density determined by a laboratory test. The most commonly used laboratory test is the Proctor compaction test, and there are two different methods in obtaining the maximum density. They are the standard Proctor and modified Proctor tests, the modified Proctor is more commonly used. However, the standard Proctor may still be the reference for small dams.

课文选自：Das B M. Principles of geotechnical engineering[M]. Boston: Cengage learning, 2006.

阅读选自：McCarthy D F. Essentials of soil mechanics and foundations[M]. Reston: Reston Publishing Company, 2007.

Lesson 9 :
Project Management

Project management may be defined as "the art and science of coordinating people, equipment, materials, money and schedules to complete a specified project on time and within approved cost". Civil engineering project management is a set of techniques and methodologies used for managing the construction or repair project of bridges, sewage systems, roads, and other civil engineering projects. The techniques used are often refined project management skills, due to the high level of complexity and low error tolerance on these projects.

For a civil engineering project, the construction phase is of vital importance because the quality of the completed project is highly dependent on the workmanship and management of construction. The quality of construction depends on the completeness and quality of the contract documents that are prepared by the designer and three other factors: laborers who have the skills necessary to produce the work, field supervisors who have the ability to coordinate the numerous activities that are required to construct the project in the field, and the quality of materials that are used for construction of the project. Skilled laborers and effective management of the skilled laborers are both required to achieve a quality project.

The construction phase is important also because a majority of the total project budget is expended during construction. The design costs for a project generally range from 7% to 12%. Using a 10% medium value, then 90% of the cost of a project is expended during construction. Thus, a 15% variation in design costs may impact the project by only 1.5%, whereas a 15% variation in construction costs may impact the project by 13.5%.

Similar to costs, the time required to build a project is always disproportionally greater than the time required to design it. Most owners have a need for use of their projects at the earliest possible date, therefore, any delay from a planned completion date can cause significant problems for both the owner and contractor. Due to the risks that are inherent in construction, and the many tasks that must be performed, the construction contractor must carefully plan, schedule and manage the project in the most efficient manner.

The objective during the construction phase is to build the project in accordance with the plans and specifications, within budget and on schedule. To achieve this objective there are three assumptions as shown in the following table.

Table 9-1

	Assumptions for construction phase
Scope	The design plans and specifications contain no errors and meet the owner' s requirements and appropriate codes and standards.
Budget	The budget is acceptable, that is, it is what the owner can afford and what the contractor can build it for, with a reasonable profit.
Schedule	The schedule is reasonable, that is, short enough to finish when he owner needs it and long enough for the contractor to do the work.

Although the assumptions are reasonable, there are often variations due to the nature of construction work. A project is a single, non-repetitive enterprise. Because each project is unique, its outcome can never be predicted with absolute confidence. To construct a project the owner generally assigns a contract to a contractor who provides all labor, equipment, material and construction services to fulfill the requirements of the plans and specifications. This requires simultaneously coordinating many tasks and operations, interpreting drawings, and contending with adverse weather conditions.

It is difficult for some individuals to acknowledge the fact that plans and specifications do have errors. The preparation of a design requires many individuals who must perform design calculations, coordinate related work, and produce many sheets of drawings that have elevations, sections, details and dimensions. Although every designer strives to achieve a flawless set of plans and specifications, this is rarely achieved.

The owner generally accepts and approves the contract documents before commencing construction. However, the plans and specifications don't always represent what the owner wants. The interest of some owners, particularly non-profit organizations or public agencies, is represented by individuals who are members of a board of trustees, board of directors, or a commission. These individuals generally have a background in business enterprises and/or professional occupations with little or no knowledge of project work or interpretation of drawings. Thus, they may approve the selection of a material or configuration of a project without fully understanding what it looks like until it is being installed during construction.

Serious problems can arise for both the owner and contractor if the contractor submits a bid price that is lower than required to build the project, with a reasonable profit. A contractor that has underbid a project can also cause significant problems for the design organization. A construction company is a business enterprise that must achieve a profit to continue operations. A careful evaluation of each contractor's bid is necessary before the award of a construction contract, because if a project is underbid by the construction contractor, the management of the project will be difficult regardless of the ability of the individuals that are involved.

Conditions that alter the project budget and schedule can arise, such as changes desired by the owner during construction, modifications of design, or differing site conditions. To reduce the impact of these conditions, there should be a reasonable contingency to allow for these types of variations that can adversely affect the project budget and schedule.

Sufficient time must be allowed for contractors to perform their work. If a reasonable time is not allowed, the productivity of workers and quality of the project will be adversely affected. There are always conditions that arise during construction and that can disrupt the continuous flow of work, such as weather, delivery of materials, clarification of questions related to design, and inspection. The contractor must plan and anticipate the total requirements of the project and develop a schedule to allow for a reasonable variation of time that may happen in the construction process.

The project manager must contend with problems as described above. He or she must always be alert to these situations and must continually plan, alter and coordinate the project to handle the situations as they arise.

1. Phrases and Expressions

methodology |ˌmeθəˈdɑːlədʒi| *n.*（进行教学、研究的）一套方法
variation |ˌveriˈeɪʃn| *n.*（等级或数量的）变动
nature |ˈneɪtʃə (r)| *n.* 性质
enterprise |ˈentərpraɪz| *n.*（困难的或复杂的）事业；项目
simultaneously |ˌsaɪmlˈteɪniəsli| *adv.* 同时
interpret |ɪnˈtɜːrprət| *vt.* 理解
adverse |ədˈvɜːrs| *adj.* 不利的
elevation |ˌelɪˈveɪʃn| *n.*（建筑物）立面；立面图；立视图
commission |kəˈmɪʃn| *n.* 委员会
configuration |kənˌfɪgjəˈreɪʃn| *n.* 安排；格局；布局
award |əˈwɔːrd| *n.* 给予
contingency |kənˈtɪndʒənsi| *n.* 可能性；不测事件
contend with 处理；对付；解决

2. Please translate the following sentences into Chinese

(1) Civil engineering project management is a set of techniques and methodologies used for managing the construction or repair project of bridges, sewage systems, roads, and other civil engineering projects.

(2) The quality of construction depends on the completeness and the quality of the contract documents that are prepared by the designer and the three other factors: laborers who have the skills necessary to produce the work, field supervisors who have the ability to coordinate the numerous activities that are required to construct the project in the field, and the quality of materials that are used for construction of the project.

(3) Due to the risks that are inherent in construction, and the many tasks that must be performed, the construction contractor must carefully plan, schedule and manage the project in the most efficient manner.

(4) The preparation of a design requires many individuals who must perform design calculation, coordinate related work, and produce many sheets of drawings that have elevations, sections, details and dimensions.

(5) Conditions that alter the project budget and schedule can arise, such as changes desired by the owner during construction, modifications of design, or differing site conditions.

3. Please translate the following sentences into English

（1）虽然大公司所取得的专业方面的进步可能来源于技术或管理知识的培训，但根据我们的经验，那些注重发展管理的公司最终会赢得更高的赞誉和认可，因为好的管理能确保工程在财政预算内按期完成，并达到客户满意。

（2）当工程项目在异地，尤其是在国外的时候，并且客户要求在当地完成设计，公司就很有必要请一名项目经理，既要精通工程技术，有很好的判断力，同时也要谙熟合同管理，以便在必要的时候就工程范围和工期变动问题随时跟客户进行磋商。

（3）项目管理是将知识、技能、工具和技术应用到项目活动中从而满足项目的要求。

（4）项目的建造包括诸多细节和各种各样复杂的关系，例如处理和业主、建筑师、工程师、总承包商、专业承包商、厂商、材料经销商、设备经销商之间的关系，以及和政府各部门和各机构、劳动者等人的关系。

（5）设计者必须对建筑工程中的任何一项已完成的工作极为关注，这样对于现场条件、材料和施工工作的任何改变都能做出评估，并且如果必要的话，还可以及时改正和改进。

4. Extension

American Society of Civil Engineers (ASCE)：美国土木工程师协会
general contractor *n.* 总承包商
subcontractors |sʌbˈkɑːntræktəz| *n.* 分包商

material dealer　*n.* 材料经销商
equipment distributor　*n.* 设备批发商
site conditions　*n.*（建筑）工地条件
residential construction　*n.* 住宅建设
biding and quotations　*n.* 投标和报价
critical path method　*n.* 关键路径法
construction site　*n.* 施工现场
commencement date　开工时间
time for completion　竣工时间
program (me)　*n.* 进度计划
procurement　|prəˈkjʊrmənt|　*n.* 采购
contract agreement　合同协议书
letter of acceptance　中标函
letter of tender　投标函
specification　|ˌspesɪfɪˈkeɪʃn|　规范
drawing　|ˈdrɔːɪŋ|　图纸
lump-sum contract　总价合同
unit-price contract　单价合同
cost-plus-a-fixed-fee contract　成本加酬金合同
tests on completion　竣工检验
civil works　土建，土木工程

5. Reading material

Typical Building Construction Projects

Building construction projects range from small office buildings valued at less than \$1 million to large skyscrapers valued well over \$100 million. The small structures might be built by a contracting firm having less than five full-time employees, whereas only the largest of contractors would typically be engaged to construct the multimillion dollar projects. Independent of the dollar-size consideration, and independent of the fact that building construction consists of a wide range of type of projects, certain types of building construction projects are most prevalent. Included in these are the following:

Office Buildings	Factories	Hospitals
Shopping Complexes	Schools	Municipal Buildings

Office buildings, shopping complexes and factories are typically built for private owners. Few of any public funds would likely be involved. While the project owner varies in terms of knowledge or skills regarding the construction industry, typically the owner of these types of projects plays a more active role in the building process than in the other types of projects listed. For example, a factory/manufacturer may have in-house designers that interact with the engaged building contractor.

External parties to include sureties, banks and government agencies play a less active role in the construction of office buildings, shopping complexes and factories than they do in most of the other types of projects listed. For example, the owners of an office building may waive the requirement of a performance bond from the engaged contractor if the owner views the firm financially sound.

There are exceptions to the fact that office buildings, shopping complexes and factories are built by private owners. For example, the General Services Administration (GSA) builds several office buildings as a branch of the federal government. When this is the case the process is more affected by third-party requirements.

School, hospital and municipal projects are characterized by a project owner that is represented by a board. For example, the owner interests for a hospital project may be represented by a board of six individuals. Often these board members are not part of the construction industry. They often are professional people to include doctors, lawyers, politicians and so forth. The end result is that the project owner becomes relatively passive in the building process. In effect the project owner becomes very dependent on the trust and knowledge of the project designer and contractor. Given this type of project owner, it is somewhat common for the contractor to be delayed owing to the fact that the owner may be indecisive in regard to decisions, e.g., selection of finish materials.

The last three types of projects listed are more characterized as being affected by external parties and being dependent on the use of public funds. For example, owing to the Miller Act, the owner will require the contractor to obtain various bonds to include a performance and material and labor bond.

There are more types of building construction projects other than the six types listed. However the typical building contractor would be most frequently involved in the types listed.

课文选自：杨金才，肖飞，李明月．土木工程英语 [M]．北京：外语教学与研究出版社，2019.

阅读选自：苏小卒．土木工程专业英语 [M]．上海：同济大学出版社，2018.